Advance Praise for "When a Butterfly Sneezes"

"In this new century, education will increasingly mean the ability to think systemically—in terms of relationships, patterns, contexts, and processes. Linda Booth Sweeney offers us an inspiring guide book to this kind of education, exploring the meaning of ancient 'systemic' folk wisdom as it speaks to us in timeless stories. This is an important and truly delightful book."

—Fritjof Capra, author, ***The Web of Life***

"Since the time of Aesop, people have used stories to illustrate important lessons about the intricate natural and social systems in which we live. Today, it is more important than ever that we all learn about the dynamics of complex systems, and from the earliest age. ***When a Butterfly Sneezes*** is a wonderful resource to do just that—rich with important systems lessons and guidance to help teachers and parents use these powerful tales to learn about our increasingly interconnected world with kids of all ages."

—John Sterman, J. Spencer Standish Professor of Management and director of the System Dynamics Group at MIT

"Systems thinking and stories are both valuable ways of understanding relationships among the seemingly disconnected parts of our experience. Linda Booth Sweeney's smart, funny, vivid book shows teachers and parents how they can use story books to help children of all ages develop their natural capability to think about systems. Great reading for everybody who cares about developing (and being) wise citizens in our global community."

—Stone Wiske, director of the Educational Technology Center, Harvard Graduate School of Education; editor of ***Teaching for Understanding***

"Bravo to Linda Booth Sweeney for making systems thinking accessible to 'kids big and small.' It's obvious by now that facts and figures are not enough to prepare us for this rapidly changing world. We need the tools and the imagination to see relationships between things, and to see the ways they interact to shape our lives and our society. Recognizing that there's nothing like stories to train the imagination, Linda Booth Sweeney provides a guide to a good supply of favorites, along with guidelines for detecting and demonstrating the systems principles at play. This refreshing book will open not only children's minds, but also the minds of adults who work with them."

—Joanna Macy, author of ***The Dharma of Living Systems*** and ***Coming Back to Life***

"All the world's ethical traditions have their roots in stories. Archetypal stories teach us to see the world in unique ways—as an interdependent system where today's gains may presage tomorrow's disappointments, where doing what makes sense for me may eventually make everything worse for us. Many ethical failings of our world today rest in the declining role of such stories in raising our children. At the same time, research indicates that children have innate skills in systems thinking and in seeing interdependencies, most of which go undeveloped or are even actively discouraged in school. Thanks to this wonderful little guidebook, parents can now join the growing numbers of educators in developing children's innate capacities for systems thinking. Help your children discover the systems lessons in many of their favorite stories and explore together a way of thinking about our lives as interrelated with one another and with nature."

—Peter Senge, best-selling author of ***The Fifth Discipline: The Art and Practice of the Learning Organization***

"Systems thinking is a distinctly modern concept with ancient roots. Linda Booth Sweeney's ingenious ***When a Butterfly Sneezes*** tells parents how to get children started on this important area of development—and parents might learn a bit from it too!"

—David Perkins, Harvard Graduate School of Education professor, author of ***Outsmarting IQ***

"How will humans learn to live peacefully with one another and the living earth in a new millennium already shadowed by violence? Begin with the children! Begin with storytelling, that oldest of human teaching devices, to empower the already active curiosity of children about 'how things work.' Linda Booth Sweeney's wonderful little book is soundly based on the general systems body of thought that has evolved over the past half-century. Delightfully, it opens the way for any adult who interacts with children to enter a profound discovery process—one that cuts through the cliches of modernity and conventional thinking about power and the use of force. Great for home-schoolers, but great for classrooms and playgrounds, too."

—Elise Boulding, professor emeritus of sociology, Dartmouth College; author of ***Cultures of Peace: The Hidden Side of History***

When a Butterfly Sneezes

by Linda Booth Sweeney

A Guide for Helping Kids Explore Interconnections in Our World Through Favorite Stories

Pegasus Communications, Inc.
Waltham

When a Butterfly Sneezes: A Guide for Helping Kids Explore Interconnections in Our World Through Favorite Stories
by Linda Booth Sweeney

Credits and permission to reprint book cover images can be found on p. 118.

Library of Congress Cataloging-in-Publication Data

Sweeney, Linda Booth, 1963–
When a butterfly sneezes: a guide for helping kids explore interconnections in our world through favorite stories / by Linda Booth Sweeney.—1st ed.
p. cm.
Includes bibliographical references.
ISBN 1-883823-52-8 (pbk.)
1. System theory—Study and teaching. I. Title.

Q295 .S92 2001
003—dc21
00-052855

Editor: Kellie Wardman O'Reilly
Design and Production: Judy Walker
Printed in the United States of America; first printing, January 2001.

Volume Discount Schedule: *When a Butterfly Sneezes*
1–4 copies $14.95 each • 50–149 copies $13.46 each
5–19 copies $11.96 each • 150–299 copies $10.47 each
20–49 copies $8.97 each • 300+ copies $7.48 each

Prices and discounts are subject to change without notice.

Pegasus Communications, Inc. is dedicated to providing resources that help people explore, understand, articulate, and address the challenges they face in managing the complexities of a changing world. Since 1989, Pegasus has worked to build a community of systems thinking and organizational development practitioners through newsletters, books, audio and video tapes, and its annual *Systems Thinking in Action®* Conference and other events. For more information, contact us at:

Orders and Payments Offices:
PO Box 2241
Williston, VT 05495 USA
Phone: (800) 272-0945 / (802) 862-0095
Fax: (802) 864-7626
Email: customerservice@pegasuscom.com

Editorial and Administrative Offices:
One Moody Street
Waltham, MA 02453-5339
Phone: (781) 398-9700
Fax: (781) 894-7175
Email: info@pegasuscom.com

www.pegasuscom.com

5381

06 05 04 03 02 01 10 9 8 7 6 5 4 3 2 1

For John and Jack

Table of Contents

In Appreciation

The seedlings for this book were planted during my first Outward Bound trip over ten years ago. While on that two-week trip, I spent four days and nights in the canyon-lands of Utah, "on solo"—no food, no telephones, no distractions—just myself and a dog-eared copy of Theillard de Chardin's ***The Divine Milieu.*** Quite a challenge for a student coming from the "24-7" pace of New York City.

The experience threw me, bare-souled and wide-eyed, into the arms of Mother Nature. I came to appreciate the sensibility of her cycles, relishing the arrivals of dawn and dusk and developing a deep awareness of the well-balanced ecosystems nested within the desert and the surrounding canyons. I didn't *study* systems during that trip, I *lived* them. I got "it" in my bones—that sense of balanced interrelationships and an implicate order among the clouds, moon, the sun, the insects, the earth, the water, and myself.

When I returned from "solo" and reunited with the crew, I noticed that we shared our experiences by telling stories. We circled around warm campfires at night and life stories poured out. As we relaxed under the star-lit sky, long-forgotten childhood tales of magic and fantasy sprang to life. (I came back to this connection between natural systems and stories as I began to write this book.)

After returning to work at a big advertising agency in the city, I realized I had internalized a new rhythm and a new way of looking at the world around me. And I had a rekindled love for listening to and telling stories. What was I to do now? How could I apply the lessons I was learning from natural systems to everyday life?

I soon left advertising, began to work for Outward Bound, and around the same time, discovered ***The Fifth Discipline: The Art and Practice of the Learning Organization,*** by Peter Senge. Here was someone who was applying insights gained from natural systems to organizational life. With rare, thunderbolt clarity, I knew that this was my next step! With the generous help of Bette Gardner (and some serendipity), I met Peter and eventually left New York to pursue a graduate degree at Harvard (after receiving more encouragement from the New York City Outward Bound Center). During graduate school, I began to conduct research with Peter and other colleagues at MIT's Organizational Learning Center (now The Society for Organizational Learning). Despite my somewhat untraditional background, Peter and others at the OLC treated me as a serious colleague, and provided me with encouragement, mentoring, and support, for which I am deeply grateful.

Many other people have contributed in many ways to water the seedling that was planted during that Outward Bound trip:

- Thank you to those who have been "systems thinking" mentors, coaches, and sources of inspiration: Stephanie Ryan, Don Seville, Dennis Meadows, Colleen Lannon, Joanna Macy, Fritjof Capra, John Sterman, Donella Meadows, Daniel Kim, Sara Schley, Rick Karash, George Richardson, Russell Ackoff, and Rogelio Oliva.
- Sheryl Erickson and I were standing on the top of a mountain when we first talked about ways to make systems thinking ideas more accessible. At the time, our ideas ranged from experiential exercises to stories to improvisation. With that conversation, the idea for ***The Systems Thinking Playbook*** series was born, and was eventually realized (with the help of Turning Point and Terri Seever). I will always be thankful for Sheryl's vision and gentle spirit.
- The idea for ***When a Butterfly Sneezes*** came in 1998, after I ran a pilot study with local fifth-grade students as part of doctoral work at Harvard's Graduate School of Education. I would like to thank the students at Longfellow Middle School in Cambridge, their teacher Charlene Morrison, and Virginia Clough for giving me the opportunity to explore this idea. And thanks to Tina Grotzer, David Perkins, Terry Tivnan, and Chris Unger of Harvard for their supervision of that study. Their thoughtful

comments encouraged me to pursue this idea further.

- Thank you as well to the many people who, over the past three years, who have listened to my budding idea of learning about systems through stories, who gave encouraging responses and help with book suggestions and other references. These people include: Tim Joy, Niall Palfryman, Philip Ramsey, Tim Lucas, Jody House, Clelia Scott, Tracy Benson, Deb Lyneis, Larry Weathers, Lees Stuntz, Mary Scheetz, Richard Turnock, Miguel Nussbaum, Shelley Bruder, Nan Gill, Nan Lux (and others on the system dynamics K–12 list-serv), and the librarians at the Cambridge Public Library, Boudreau branch, and Judy Greenfield at the Rye (New York) public library. Thanks also to those on the NESCI list-serv, who helped me make sense of some challenging chaos theory concepts, especially Arthur Battram, Gary Nelson, Chris Hancock, and John McCrone.
- To those who made this book come alive—Kellie Wardman O'Reilly, Laurie Johnson, and Ginny Wiley of Pegasus Communications—thank you for your vision, sense of humor, and your special brand of support, enthusiasm, and encouragement. Thank you especially to Kellie, who, like me, as the mother of a toddler, would exchange book drafts with me long before "the boys" got up in the morning. She has shepherded this book from "good idea" to something we all hope will be of real value to parents and educators. This book is a testament to her talent and care. And to Judy Walker, the designer of the cover and book, for your commitment to making this visually exciting and appealing.
- Thank you to those who read final drafts and provided invaluable comments: Dawna Markova, Elise Boulding, Joanna Macy, Mitchel Resnick, Mary Ellen Gonzales, John Sterman, Paula Underwood, Don Robadue, Mary Catherine Bateson, David Eddy Spicer, Carol Ann Zulauf, Jennifer Berger, Janice Kowalczyk, John Marcus and Susan Prout. And a very special

thank you to Diane Cory for her numerous readings and her gentle, incisive, and tenacious pursuit to bring "Linda" back into the manuscript.

- Thank you to Sarita Chawla, Ronita Johnson, Anne Doscher, Kristin Cobble, Andrea Dyer, Teresa Ruelas, Stephanie Ryan, Gisella Wendling, and Peggy Sebera for the cherished circle of friendship and the place to bring crazy "what if's."
- My infinite appreciation to my parents, who lovingly read draft after draft, encouraged me to write, and gave me that look that said, "you are something!" Thank you to my nephew Bradley (and to my brother Toby) for your illustrations of ***If You Give a Mouse,*** and to the Sweeney family for understanding when I'd steal away to write during some of our family visits.
- Finally, "mille gratzies" to my husband John, for believing in this book and me, and for moving heaven and earth to make time and space for me write. And to my son Jack, who as our "sunshine boy" has provided more inspiration for writing this book than I could ever have imagined.

Linda Booth Sweeney
Cambridge, Massachusetts

Foreword

by Dawna Markova

Who first taught you to see beauty in the world? Who taught you to notice the things you'd never noticed before—a hummingbird hovering in the still center of its own motion? Who gave you the gift of knowing that you belong, the gift of wonder that opens your mind again and again to unlimited horizons?

Perhaps even more important, how do we, as parents or educators, hold a young person's mind as the sky holds the birds? How do we teach children to think for themselves, to think with open minds, to think in such a way that they take responsibility for their own lives?

For the past 300 years, many teachers and parents have instructed children from the outside in, assuming that a child's ignorance can be filled with our expertise. We have decided what they should learn and how they should learn it. We have ignored their own awareness of themselves and the effect of their actions on the world. We have ignored their own natural gift of wonder.

Parenting and teaching this way is hard work, and seems to result in adults who feel both overwhelmed and isolated. When I stood at the top of our driveway teaching my son to ride a two-wheeled bicycle without training wheels, every muscle in my body was as tight as a violin string. "Watch out, David, you're going to hit that tree! Not so fast—you'll crash. Slow down; you're going all over the place. Listen to me and slow down or else! OK, you've had it. That's enough for today."

This is how most of us learned to doubt ourselves, our own capacities, and the world. And most of us interfere with our children's learning in the same way as we interfere with our own.

There is a new story in the air—a story that entices us toward a kind of teaching that is more worthwhile than instruction simply leading to better grades and higher test scores. My son, a few decades later, was living in it. He stood at the top of a hill and helped his mother learn to make turns on a ski slope. "OK, now Mom, let's experiment. What do you think will happen if you start going across the hill and if you press more on the right inside edge of your feet than on the left? And what would happen if you press more on the left? How about if you press your toes into the boots and then the heels?"

"But David, can't you just tell me the right way to do this?"

"I don't know the right way, Mom, but your skis and the snow will tell you. They'll write patterns in the snow the same way your hand and the pen write letters on a page. Let's just be curious."

He stood there relaxing in the blazing Utah sun, flagrantly interested in the outcome of this experiment. He hadn't given me a single instruction or warning. All he offered were questions that formed a conversation of discovery and the excitement of having a co-conspirator in the unknown.

In the past three and a half decades, I have served as a thinking partner to thousands of children of all ages, graduate students, parents, teachers, healthcare providers, families, couples, and corporate executives. They have come to me seeking answers, cures, and solutions. I realize now that they all brought hands, hearts, and minds full of the wrong questions. I realize now that for at least half of those years, I was more committed to instructing than learning.

How would our lives be different if we asked questions such as "How can I help my child love to learn?" Questions that lead to conversations of discovery, to experiments and explorations, questions that open all of our minds? Children naturally wonder their way across the abyss of the unknown with such questions. "Daddy, why can't something just grow forever?" Their questions are thought experiments, the very yeast that helps their minds rise and risk the reach forward and through the mysteries they encounter.

What if we, as thinking partners to our children, learned how to enter into such conversations of discovery with them? What if you and I learned to ask questions to which we could not possibly know the answers? Questions that help our children think for themselves are those that increase their awareness of what they are doing, what their motivations are, what effect their actions have on others and the world, and what the consequences and after-effects of those actions might be.

This engaging guidebook that Linda Booth Sweeney has crafted is like the door to a precious sacred temple. ***When a Butterfly Sneezes*** invites us to step inside and be refreshed, to be less of who we think we are and more a part of everything. It reminds us that wisdom is not about bits and pieces, but about relationships, and about the compassion that comes when we realize our deep relatedness. The book guides us in an exploration of stories, stories that pass on profound truths and cultivate everyday wisdom. It suggests that if we share different stories with our children and help them think through them in a different way, perhaps the stories they tell themselves about their own capabilities and capacity to make things happen in the world will also be different. It leads us to wonder whether helping children discover meaning in a story might also help them discover meaning and purpose in their own lives.

As you leave this little temple and re-enter your ordinary life, you are pulled forward by luscious questions: How do we teach our children to love and serve the gift of life? How do we help them cultivate wisdom in an age that barters bits and pieces of information? How do we help them think all the way through a problem, thinking deeper and wider even as technology insists they think faster? How do we help them navigate on the edge of chaos?

Children should not fail. If they do, it is we who have failed them. ***When a Butterfly Sneezes*** brings us a great gift—guidance in fostering the love of learning in our children's lives, and in creating the conditions that help them understand how they belong to the mystery of an interconnected and interdependent world.

—Dawna Markova, Ph.D., co-editor of ***Random Acts of Kindness*** and ***Kids' Random Acts of Kindness,*** and author of ***How Your Child Is Smart, Learning Unlimited,*** and ***I Will Not Die an Unlived Life***

From the Author

If you've picked up this book, you're probably a parent or an educator—or maybe even both. I wrote ***When a Butterfly Sneezes*** for everyone who wants to find a fun and memorable way to help kids see and understand the world of systems all around us. Our bodies, families, our schools, our communities—these are all systems. By understanding how they work, through a field known as *systems thinking*, we can all deal more effectively with the increasing complexities of everyday life.

Why learn about systems through *stories?* Stories are fun and memorable. And, it is often through fairy tales, myths, and stories of all kinds that we pass on profound and subtle wisdom to future generations. Finally, adults ("big kids") can learn just as much from stories as kids can.

This volume describes 12 favorite tales from around the world that offer powerful lessons about life. Most are picture books and so are suitable for even the youngest of readers. Many are likely already on your bookshelf! These stories were generally written for the 4–8-year-old age range. To help you sort through them, I've compiled their descriptions by the target age groups—younger readers first and then older readers. Volume II will describe chapter books and books written primarily for older readers.

In a way, children understand the importance of story more than grown-ups do. As adults, we're taught to put away the fantasies of our childhood and to be "mature"—that is, rational and level-headed. As we store childhood stories away, we may put our imaginations and sense of wonder into the back of the closet as well. As you'll see, systems thinking requires a healthy dose of imagination. So, my first hope is that this book will help to "restory" all of us grown-ups—reigniting our sense of imagination and wonder, regardless of our age. If you're a parent, I hope this book will enhance your everyday reading sessions with your kids. For educators, I hope the book will complement the teaching methods and subject matter you're already using. After reading this book, *both* parents and educators should be able to select and interpret additional stories from a systems thinking viewpoint as well.

If you're familiar with systems thinking already, you may want to scan the chart on pp. 10–11 to find stories that bring out particular concepts of interest to you. If you're new to these ideas, I'd suggest you begin by reading Part I—Systems Thinking: A Means to Understanding Our Complex World (p. 19), as a way to familiarize yourself with basic systems thinking terms and concepts.

I have an additional hope, this one for children's-book authors: May they write more stories like the ones described in this book!

Overview of the Stories' Systems Thinking Concep

The Stories	Page	Simple Intercon-nected-ness	Circular Feedback	Time Horizons and Delays	Uninten Conse-quence
If You Give a Mouse a Cookie (Numeroff)	45	x	x	x	x
The Old Ladies Who Liked Cats (Greene)	51	x	x	x	x
The Cat in the Hat Comes Back (Seuss)	56	x			x
Once a Mouse: A Fable Cut in Wood (Brown)	61	x			x
The Sneetches and Other Stories (Seuss)	64	x	x		x
Anno's Magic Seeds (Anno)	68		x		
Zoom (Banyai)	72				
A River Ran Wild (Cherry)	77	x	x		
The Butter Battle Book (Seuss)	81	x	x		
Tree of Life: The World of the African Baobab (Bash)	85	x	x	x	
The Lorax (Seuss)	88	x	x	x	x
Who Speaks for Wolf? A Native American Learning Story (Underwood)	93	x			x

ponential owth	Escalation	Levels of Perspective	Nested Systems	Structure Drives Behavior	Limits to Success	Fixes That Fail	Shifting the Burden
						x	
				x			x
x					x		
		x	x				
			x		x		
	x			x			
					x		
		x					

Why "When a Butterfly Sneezes"?

Can the flap of a butterfly's wings (or worse, a butterfly's sneeze!) in Brazil actually affect a tornado in Texas?

By answering "yes" to this question in 1972, Dr. Edward Lorenz, a research meteorologist, wreaked havoc with some of our basic assumptions about cause and effect. He raised our awareness that tiny differences in a complex system (such as a butterfly flapping its wings) can produce large and unanticipated effects, such as changing weather patterns a thousand miles away. While this phenomenon may sound apocryphal, research in chaos and complexity science seems to be pointing more and more to a world that is interconnected in many, many ways that we have not begun to understand.[1]

We don't have to look far to see that our children are growing up in a world that is more and more complex, fast changing, and interconnected. With new information technologies and more accessible media, communications that used to take a day, a week, or a month to reach a remote part of the world can now move the same distance in just seconds with the press of a button.

Similarly, the world's interconnectedness is becoming more and more evident. When Mount Saint Helen's erupted in the western United States, the rest of the world saw spectacular sunsets for the following year as the sun reflected off minute dust particles floating in the atmosphere. As fear of fallout from the Chernobyl nuclear plant traveled with the prevailing winds, the problem became Eastern Europe's and the rest of the world's, not just the USSR's.

As the "Butterfly Effect" demonstrates, our world is more global than ever. No matter where our kids grow up, their lives influence—and are influenced by—the lives of people everywhere else on the planet. Clearly, we want our children to understand these interconnections—so they can more effectively work within this web of relationships. As the parent of a bright, inquisitive toddler, I frequently wonder how he will learn to live in a world where the results of a poor corn crop in Iowa, U.S.A., affect what a little boy might have to eat in the African savanna.

From playground fights in grade school to "homework burnout" in high school, from virus outbreaks to boom-and-bust markets they'll hear about in later years, our children face all sorts of situations throughout their lives that demand their understanding and problem-solving skills. As parents and educators, we want to help them avoid getting swept up in or hurt by these sorts of events. Ideally, we want them to be able to see the systems they are embedded in, to understand why troubling things happen, and to figure out what they can do about them. As the Butterfly Effect might suggest, we want them to adopt a mindset that they are part of a larger system, *and* that they can make a difference even as one person in that world.

So, how can we help our kids grasp these realities and move into adulthood prepared to deal with them? One way to do this is to teach our children to live not just from moment to moment but with an understanding of how problems come about and how new challenges might unfold in the future. This means questioning overly simple explanations of events, looking for patterns in how things happen, experimenting, and even redesigning systems so that they work better for them.

One way for you to help children to develop these life skills is to share ideas and tools from the field of systems thinking. This way of thinking helps us see how the many social systems in the world around us—from families and neighborhoods to global economies and governments—actually work. It also helps us understand how the cause-and-effect connections among the parts that make up these systems influence the events we experience in our day-to-day lives. In particular, systems thinking helps us see how events can build on each other (for instance, when a playground argument between two bullies escalates into an all-out brawl) or control or counteract each other (a child's efforts to improve her grades can eventually be counteracted by fatigue and burn-out).

A systems view of the world is by no means new. More than 2,000 years ago, the Greeks were describing reality in terms of wholes composed of related parts. Indeed, many cultures, including Native American traditions, tend to see reality in terms of indivisible wholes, emphasizing inter-relationships and circular loops of causality. Systems thinking ideas have also been used successfully by family therapists, business managers, educators, and trainers around the world—with remarkable results.

Adopting a systems thinking stance can be an important part of successful parenting and teaching. Through systems thinking examples and stories, we can show our children how to solve, anticipate, or as systems thinker Russell Ackoff says, "dissolve" problems. We can also show them how to address the challenges facing them in their communities and the world. Systems thinking can help a child to understand how the mysterious natural and social worlds function, see how he or she contributes to trouble or creates success, and even understand the bigger picture of what parents and teachers are trying to accomplish.

Like many adults, I have an intuitive grasp of systems thinking ideas and have come to see how valuable they are in dealing with the most important challenges in everyday life. However, systems thinking was never an explicit part of my education in grade school, high school, or even college. When I did hear systems-oriented teachings, they came in the form of old adages, such as my mother's favorite "an ounce of prevention is worth a pound of cure" (her gentle reminder to anticipate unintended consequences and to try to take "the long view").

Without a doubt, systems thinking is embedded in our old teaching stories, common sense, and wisdom traditions. Yet many basic systems concepts are conspicuously missing from much of Western education. Clearly, parents and educators of young kids need some other way to share these lessons.

Stories: An Enchanting Doorway to Learn About Systems

"Story telling is the most ancient form of education."
—Joan Halifax, *The Fruitful Darkness*

Like many children, my young son Jack learns a great deal from books—when his father and I read to him and through his own paging through picture books. On some days, he consumes as many as 20 or 30 stories! His love of books has made me wonder: Why can't parents, teachers, and kids learn about systems thinking through story? But how many children's stories embody systems principles? Unfortunately, not many. Most of the kids' stories I've seen tend to describe simple plot lines that feature some sort of problem, a reaction, and a quick resolution; that is, A causes B, and B causes C. End of story. For instance, remember the story "I Know an Old Lady," about the old lady who swallowed a fly?

I know an old lady who swallowed a spider
That wriggled and wriggled and tickled inside her.
She swallowed the spider to catch the fly.
But I don't know why she swallowed the fly!
I guess she'll die!

The old lady goes on to swallow a mouse, a cat, a dog, a cow, and, finally, a horse—all to catch a pesky little fly. What happens when she swallows the horse? "She dies, of course!" While I'm delighted that Jack can repeat the rhymes from this darkly humorous story, I can also see that it is essentially a linear tale made up of a long chain of events. It doesn't teach anything about the relationships between the elements in the story or the fact that events can circle back to either amplify or lessen each other. Neither does it help kids look for patterns in events or root causes of problems. Indeed, most children's stories (particularly in Western cultures) are organized along a typically linear chain of events that are made up of the following five elements:

- An initial event that sets up the problem; for example, the old lady's swallowing the fly;
- A simple reaction, often an emotional response to the initial event; e.g., the old lady gets bothered by the fly in her stomach;
- The setting of a goal, in order to do something about the problem; for instance, the old lady decides to swallow a spider so that it will eat the fly in her stomach;
- An attempt to reach the goal; e.g., she swallows the spider, but then it wriggles and tickles inside her;
- A reaction; she keeps swallowing more and more creatures to solve her worsening problem.

This chain-like story structure raises a sobering question: Do most children's stories actually encourage kids to see the world as overly simple chains of events-reactions-resolutions? Or said another way: *Are young people, particularly in Western cultures, actually "set up" by the structure of the stories they read to look for linear rather than systemic relationships?* I think the answer to this question is a resounding "yes," and that's how this book came about.

Part I

Systems Thinking: A Means to Understanding Our Complex World

"Since relationships are the essence of the living world, one would do best . . . if one spoke a language of relationships to describe it. This is what stories do. Stories . . . are the royal road to the study of relationships. What is important in a story, what is true in it, is not the plot, the things, or the people in the story, but the relationships between them."

—Fritjof Capra, *Uncommon Wisdom: Conversations with Remarkable People*

The above quote, in which natural scientist Fritjof Capra quotes systems thinker Gregory Bateson, captures the reason that we can benefit by connecting systems thinking to stories: Both act as metaphors that can help us interpret our experiences. But before we explore how to use the stories described in this book, let's take a deeper look at what systems thinking really is.

One great first step is to understand what it is *not.* Systems thinking is not *analysis.* If you're like most people, you probably had a teacher somewhere along the way who taught you that the best way to understand something was to *analyze* it—to break it down into bite-size, manageable pieces. So, for example, to write an essay, you were taught to break it down into its component parts: the introduction, the purpose, the body (with supporting facts, of course!), and the conclusion.

Many people approach lots of challenges this way, from learning how to juggle, to applying to college, to figuring out the best way to lose weight. This slice-and-dice approach is fine for some problems—for instance, when you want to organize your CD collection or understand how a clock works (you take it apart!). It's also helpful if you want to understand the basic elements of something; for instance, figuring out that water is really made up of hydrogen and oxygen atoms.

The problem arises when we use analysis *mindlessly*, assuming that the world stands still as we study it, that puzzling situations will stand still while we break them into their component pieces, and that the relationships between the pieces aren't important. As anyone who has tried to get a growing family out the door in the morning knows, problems *don't* stand still and inter-relationships *do* matter! Analysis therefore gives us a limited understanding of reality—and so it isn't the only skill kids need to handle the big challenges in their lives.

By contrast, systems thinking helps *expand* our understanding. It shows us how to:

- see the world around us in terms of wholes, rather than as single events, or "snapshots" of life;
- see and sense how the parts of systems *work together*, rather than just see the parts as a collection of unrelated pieces;
- see how the relationships between the elements in a system influence the patterns of behavior and events to which we react;
- understand that life is always moving and changing, rather than static;
- understand how one event can influence another—even if the second event occurs a long time after the first, and "far away" from the first;
- know that what we see happening around us depends on where we are in the system;
- challenge our own assumptions about how the world works (our *mental models*)—and become aware of how they limit us (what a kid might call "stinking thinking");

- think about both the long-term and the short-term impact of our and others' actions;
- ask probing questions when things don't turn out the way we planned.

We can't *abandon* analytical thinking; after all, it really is important in dealing with certain kinds of tasks or simpler problems. But if kids know how to *complement* analytical thinking with systems thinking, they'll have a much more powerful set of tools with which to approach life.

A Child's Early Steps Toward Systems Thinking

Children are actually natural systems thinkers. They start recognizing how systems work early in their lives. At about five months, a baby begins to play games with her parents. She learns to cry deliberately, and then waits to see if her doting parents will hurry over and pick her up. If they don't, she cries again, perhaps a little louder this time. These kinds of experiences give children a basic understanding of *one-way causality:* "If I cry, my parents will come and pick me up."

As the child grows into a teen, she extends her understanding of causality to her family and community: "If I stay out late, my parents will be mad, which means I may not be able to stay out late next weekend." She thus begins to learn about *mutual causality,* and to experience the nature of interdependence between herself and others, through being a member of a family, a sports team, a neighborhood, and so forth.

As she moves into adulthood, the young woman then finds a world of accelerating change, where new, faster technology and shrinking global borders collide to create an increasingly interconnected landscape. In such an environment, everything everywhere appears to be—and actually is—connected to everything else.

As adults, we all survive in this environment by trying to make sense of the phenomena we perceive around us. We then use our explanations to predict what may happen in the future. Yet our explanations often contain misconceptions about causes and outcomes and incomplete or overly simple assumptions about how the world works. When this happens, we struggle again and again with what seem like the same problems. We take actions that we think will address fundamental problems, but often they never do, or they actually make the original problem worse. How do we get off this problem-solving treadmill?

This is where systems thinking can help.

What Are Systems, Anyway?

How can we, as adults, help youngsters learn about complex systems, when many of us don't have formal training in systems thinking? One simple step is to understand the basic characteristics of systems. The following five questions can get you started.

Q: Is it a heap or a system?
Q: Is the whole greater than the sum of its parts?
Q: What's the purpose?
Q: Are the causes and effects shaped like a circle?
Q: Are we experiencing déjà vu?

Is It a Heap or a System?

Systems consist of two or more parts, but so do "heaps," such as a bowl of mixed nuts. So how do you know if you've got a plain old heap of something, or a system? Here's a basic way to tell:

With a *heap,* nothing changes if you take away or add parts. For instance, imagine that you have a bowl of nuts. What happens if you remove all the cashews or add hazelnuts? Answer: You still have just a bowl of nuts.

With a *system,* things definitely change if you take away or add parts. For example, suppose you removed the battery from your car. The car wouldn't start! A car is an example of a mechanical system. Living systems—including human systems like our families, classrooms, neighborhoods, and nations—are far more complicated. When you take away or add parts to *those,* you get some very complex changes. (Imagine what life would be like if your town's police force or waste-disposal department completely disappeared one day!)

Children and adults can learn to distinguish between heaps and systems by doing the simple exercise described above: Just ask whether anything important would change if you took away or added parts, or if the parts can operate on their own. You will find that systems always have a distinctive arrangement between their key elements; thus the wisdom behind the old saying "If you cut a cow in half, you don't get two cows." For example, if you remove a piece of fruit from a bowl, it still functions as a piece of fruit. But if you remove your hand from your body, the hand certainly won't work in the same way it used to. That's because every part of the "body" system has certain capacities that it loses when separated from its system. What makes these capacities work is the interaction among the system's parts. (This is why analysis doesn't work when you want to understand a system as a whole.)

Is the Whole Greater Than the Sum of Its Parts?

All living systems consist of a huge number of tightly connected interactions. How do these complex interconnections play out in our daily lives? We've all heard the saying, "The whole is greater than the sum of its parts." In systems thinking, this means that the many interactions among the parts in a system give rise to qualities or properties that you just can't measure merely by adding up those parts. For example, when fishermen overharvested sea perch on California's coastline, orcas—whales that had fed on the perch—began to prey on otters instead, even though they and the otters had coexisted peacefully for centuries. This is just one example of how

the whole (the ecosystem) is affected when the interactions between the parts (the animals in the ecosystem) are affected.

As another example, imagine that you not only took the battery out of your poor car; you also completely took the car apart. If you weighed all the pieces and added up the numbers, you'd know how much the entire car weighs when it's assembled correctly. But, you wouldn't know how *fast* the car goes or how comfortable a ride you'd have on a bumpy road. *Speed* and *comfort* are created by the *interactions* of the car's parts and thus are "greater than the sum" of all the car's separate parts. *Speed* is an example of what is called an "emergent property"—a property or behavior that arises only out of the interactions within a specific set of parts.

If your children play in an orchestra, act with a drama troupe, or are members of a sports team, they are learning about emergent properties all the time. They know that the all-star team is not always the best team in the league. Why? Each person on the team may have exceptional batting or throwing skills, but when you put the players together, they don't necessarily make the best *team*.

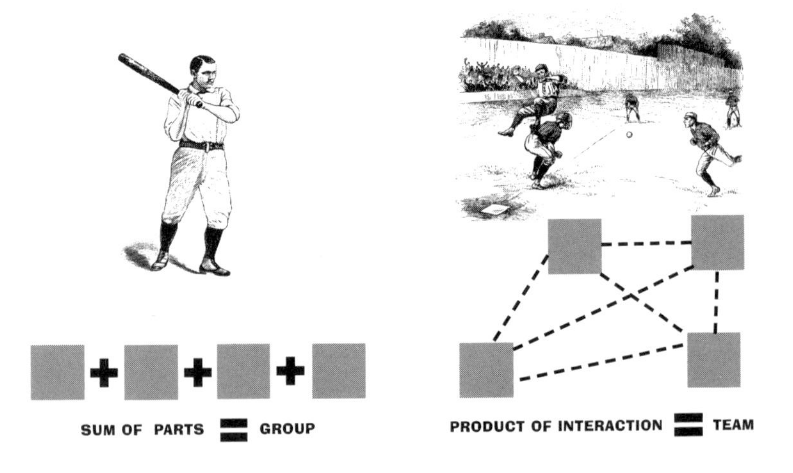

What does make a team great? It's the quality of the *interactions* among the players—which comes only from lots of practice, time spent in getting to know one another, and good old experience.

Systems thinkers have a healthy respect for emergent properties: for understanding them, intervening in them (e.g., when children on a long car trip start to get antsy), and sometimes, as in the case of team spirit, fostering them.

What's the Purpose?

Most systems have a distinct "point," or purpose in relationship to the larger system in which they are embedded. In many social systems, we see subsystems whose purposes can conflict sometimes; for instance, the teachers in your kids' school may sometimes be at cross-purposes with the guidance department, or with the administration.

The simple question we might often forget to ask is this: "What is the purpose of this system?," whether it be a clique in school, a committee on the school board, or a department in an organization. By understanding the various and sometimes conflicting purposes within a system, you can begin gaining insight into why the system functions as it does—and how you might help it function better. (For example, regular meetings between teachers and school administrators might help them clear up conflicts.)

Are the Causes and Effects Shaped Like a Circle?

We can think of cause and effect, or causality, as coming in several "shapes." The shape can be linear, as in the story of the old lady who swallowed a fly:

Many people—children and adults—assume that this is how things happen: One thing causes the next, like a set of falling dominos. In the case of dominos, this is indeed the nature of causality.[2]

Systems thinkers have another notion of causality: feedback loops, which have a circular shape. The simplest way to think of these is to imagine that one event causes another event, and that second event comes back around to influence that *first* cause. It's like this: A causes B, B causes C, and C causes A. For example, the more children who are born, the greater the population. And the greater the population, the greater number of possible parents. And the greater number of possible parents, the greater the number of births ... and so on (assuming a relatively stable environment).

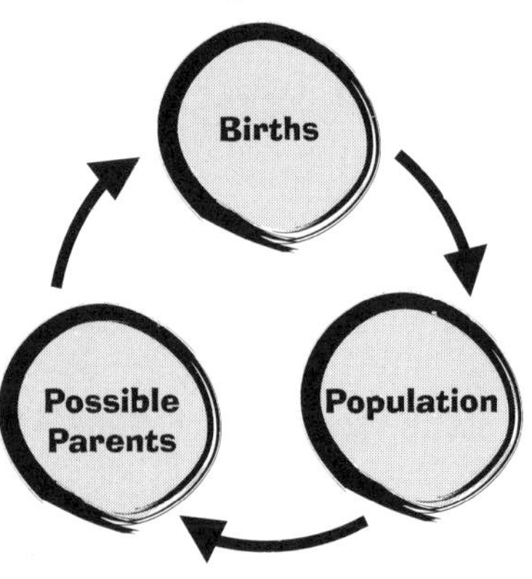

This idea is by no means new: Many indigenous cultures see the world in terms of circular causality. In a well-known speech, Black Elk, a holy man of the Oglala Sioux, once said:

> ***"Everything an Indian does is in a circle, and that is because the power of the world always works in circles, and everything tries to be round."***

How do we see these circular loops in everyday life? Here's an example: My little son Jack likes green peas (which makes his mother very happy!). When he was about one, he would eat a few peas and then, because it must have seemed like a fun thing to do, he would throw some peas on the floor. Then he'd say, "I did it!" Being the novice mother I was, I laughed, thinking that the whole thing was funny. What did he do? He then ate *more* peas, threw *more* peas, and said, even *more* happily, and proud of himself, "I did it!!!" Of course, I'd laugh even more—and then he'd throw peas even more!

This is a simple example of what systems thinkers call a *rein-*

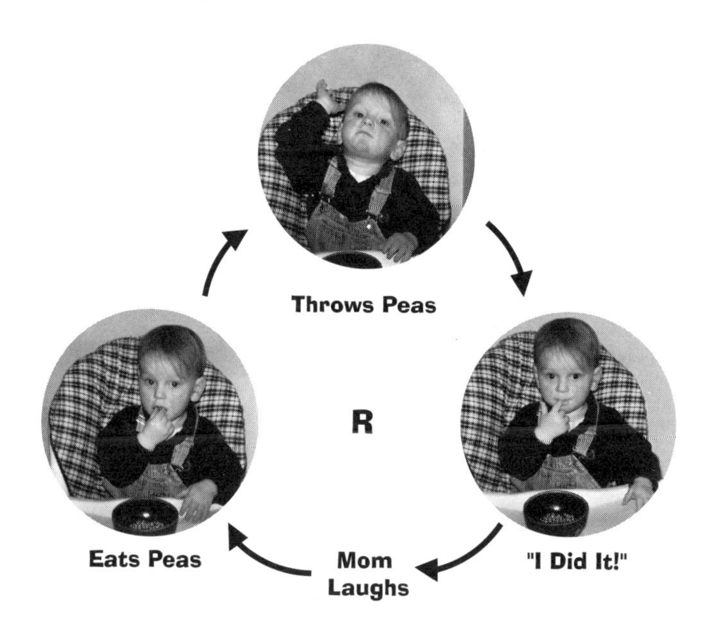

forcing feedback loop (other examples include teachers' expectations of student performance, population growth, and compound interest building in your savings account. For more information on how to draw your own feedback loops, see "Guide to Systems Thinking Diagrams" on p. 102). Reinforcing loops cause dramatic growth or collapse (like the stock-market crash in the U.S. in the 1930s). They're usually easy to spot because they're so extreme. When you hear people say things like "The situation's totally out of control" or "Things are just snowballing," chances are there are reinforcing loops at work.

Depending on the situation, a reinforcing loop can be either vicious (amplifying to make something greater) or virtuous (amplifying in the opposite direction, e.g., making something less).

But as the old systems adage goes, "Nothing grows forever." Luckily, there's another kind of feedback loop that helps to keep things under control in general. (Otherwise, Jack would be throwing more and more peas into infinity, and I'd be laughing more and more—with no end in sight to the cycle!) Systems thinkers call this other kind of loop *balancing.* Balancing loops put limits on dramatic growth and collapse, and ensure that a system fulfills its purpose.

For example, in the case of Jack's throwing peas, the balancing came from the fact that Jack eventually ran out of peas to throw. I also provided a balancing action: I quickly learned to stop laughing and changed the goal of the "game" by encouraging him to see how many peas he could stack in a bowl instead of throw on the floor.

Another simple example is your body's temperature-control system—whose purpose is to keep your body temperature at 98.6 degrees Fahrenheit.

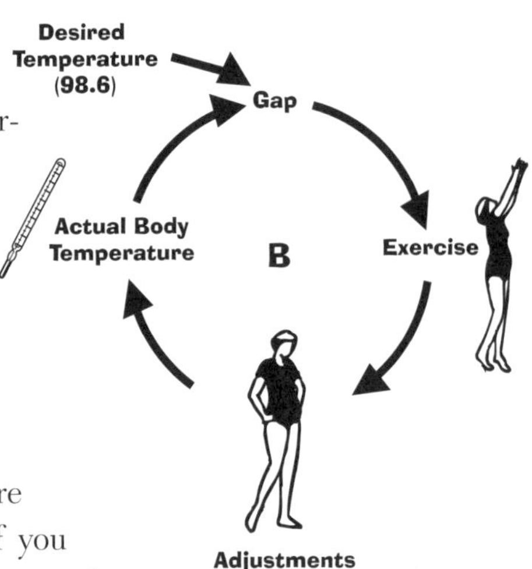

When you exercise, your body heats up. To keep your temperature close to 98.6 degrees Farenheit, your sweat glands produce perspiration. Air moving across sweaty skin creates a cooling effect that helps brings your temperature back in balance. And if you get chilled, your muscles start shivering, and the friction of your shaking muscles warms your body back up to 98.6 degrees. Balancing loops aren't as noticeable as reinforcing loops, but there are lots of them out there. Because they tend to keep things steady, we can sometimes detect them when we try to change something but get no results (e.g., when we *try* to lose weight!).

Are We Experiencing Déjà Vu?

Another interesting thing about systems is that they tend to behave in similar ways in very *different* kinds of settings. For instance, let's go back to that age-old problem of playground fights.

Counting Chickens in a Complex World

There are only two types of loops—reinforcing and balancing. The interaction (usually nonlinear) of these loops creates all dynamics that we experience in all systems. Indeed, some of the most confounding behaviors we face arise when multiple reinforcing and balancing loops interact.

Say we want to look at that age-old dilemma of chicken-and-egg populations. Assuming we have at least one rooster, the population of chickens likely goes up and down over time. But looking at the reinforcing loop between the number of chickens and the number of eggs laid will give us only part of the story. For example, suppose that in this case, the roost is actually located right next to a highway. As the chicken population grows, there is more chance of road crossings (a balancing loop that leads to fewer chickens). If this balancing loop were the only one operating (because the farmer decided to sell all the eggs), then over time, there would eventually be no more chickens left.

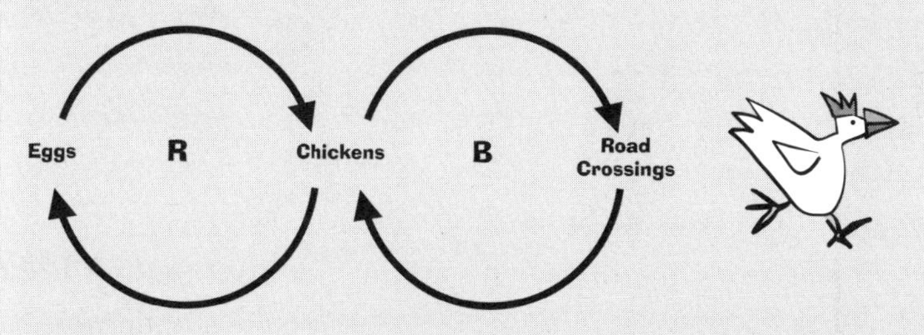

So, because most knotty problems we find ourselves trying to understand arise out of some interaction of balancing and reinforcing loops, our challenge is twofold: 1) to be aware of our own oversimplified perceptions of feedback and 2) to discover and be able to explain these interactions to others.

Source: John Sterman, *Business Dynamics: Systems Thinking and Modeling for a Complex World.* 2000. New York: Irwin/McGraw-Hill. Reprinted with permission.

One bully insults another, who then comes back with an even more inflammatory retort. The next thing you know, someone throws a punch—and an all-out brawl erupts. Now think of companies competing in the business world. One draws more customers by slashing prices. Its main competitor, concerned about being left behind, slashes *its* prices even more—prompting the first company to try to offer even lower prices. Even though these two situations look very different on the surface, both involve a build-up, or an escalation, of tensions or competitiveness.

Systems thinkers have identified a whole set of common "stories" like this—which they call "systems archetypes"—that occur in very different settings. For instance, in an archetype called "Fixes That Fail," you do something to try to solve a problem—but the problem eventually just gets worse (like drinking coffee to perk you up when you're tired, but then when bedtime comes, you're restless and don't sleep as well, and you get even *more* tired over time).

With another archetype, "Limits to Success," you get something really great going, but then it seems to level off—such as persuading more and more neighbors to help out at the annual town fair—but then suddenly seeing their help taper off. What has happened? A reinforcing loop (efforts to make the town fair successful) is connected to a balancing loop (a limit to the time, energy, or resources that folks in town can provide). If you have that "I'm spinning my wheels" feeling, you may be experiencing this archetype.[3]

The great thing about these old teaching stories is that they let you recognize—and therefore better manage—common problems that occur in lots of different situations. And they also tend to show up in some favorite stories!

Ready for More?

Still with me? Good! You've now got a basic idea of what systems are and how they behave. Want to go a little further? Here are some other tried-and-true ways to think about systems.

- "Lowering the Water Line": Seeing How a System's Structure Influences Behavior
- "Oops—I Didn't Mean to Do That!": Understanding Unintended Consequences
- The Systems Thinker's Clock: A Different Look at Time
- Ferreting out Delays
- Thinking Like a Bathtub

Lowering the Water Line: Seeing How a System's Structure Influences Behavior

One of the most profound and practical habits I've learned from systems thinking is to consciously look at a system from multiple perspectives—actually, from multiple *levels* of perspective. To use an example a teacher might face, if Johnny is late to school, he might lose his recess privilege that day. But this "event" often doesn't tell the whole story. Systems thinking tells us to stop and look below the surface, to see how the structure (the relationships between the parts in the system) drives the patterns of behaviors we see. These patterns influence the events to which we react.

Think of this idea as an iceberg, where only 10 percent of the ice is "above the water line" and 90 percent of it is beneath the surface (see p. 32). In Johnny's case, the "event" is that he is late. If we look below the water line, we might see a certain pattern of behavior; that is, Johnny is late every Tuesday and Thursday morning. If we go down one more level, by asking Johnny and his family why this is happening, we might find that on Tuesday and Thursday mornings, Johnny's mother has to drive another sibling to preschool, which often makes Johnny late for class. If the teacher and the parent actually move their conversation and exploration to this level, they might come up with any number of creative solutions to the problem.

Without a doubt, the things that most preoccupy us and are easiest to react to are events—a fire breaks out; we pull off a "miracle" to get a project done on time; the stock market jumps up or

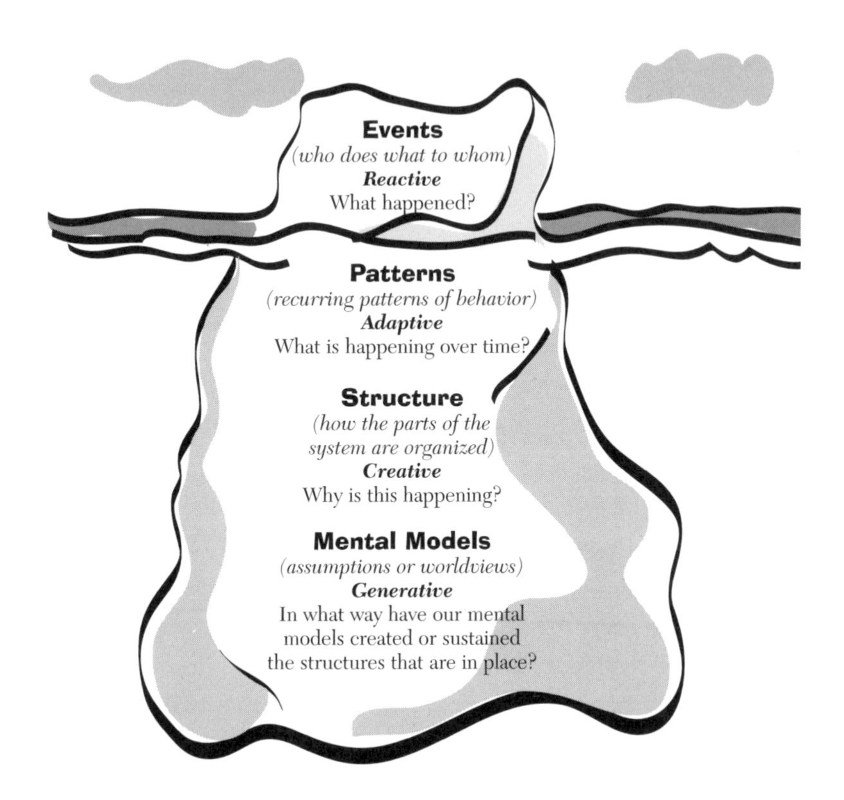

down, oil prices skyrocket. And as a general rule, not only do we tend to focus *first* on events, but we then look at the immediate actions we need to take, rather than think about how the situations fit into a larger pattern or what may be causing them in the first place.

At the same time, it's important to remember that reacting at the event level is quite appropriate in certain situations. For example, if someone is running in front of a car, shouting or even pushing them out of the way is the right thing to do! But as in my own neighborhood, if you see that people are getting hit on the same corner a few times a month, then it would be useful to "lower the water line" and look at what set of interactions might be causing the pattern of accidents. So, for example, you might discover that kids are crossing the street at a time of day when the sun is most blinding, a problem that could be addressed by providing children and adults with bright orange flags to wave as they cross the street.

"*Oops—I Didn't Mean to Do* That!": *Understanding Unintended Consequences*

Often when we have a problem, we look for a fast way to fix it. While our solution may ease the problem for a little while, it might also have some consequences that we never intended—and that even make the original problem worse eventually.

You may experience this phenomenon in your own life. For example, suppose you want to spend more time with your kids because you believe that quality time together will help make them happy. So, you play with them after work, keeping them up past their bedtimes. They get more time with you, but as an unintended consequence, they're cranky and tired the next morning—hardly the happy individuals you wanted them to be!

The more you know about systems thinking, the more you can *anticipate* unintended consequences instead of being caught off-guard by them. Sometimes, the *actions* that cause the consequences can seem relatively insignificant. For instance, suppose it's a hot and muggy Saturday afternoon, and the kids are restless. You decide to take them to the local beach to try to cool down. You pack everybody up and drive over to the lake—and discover that everyone else in town had exactly the same idea! Families are crammed towel-to-towel all along the beach, and the designated swimming area is so full that no one can actually swim. Everybody's getting annoyed—including your kids, who are even more irritable than before! You're irritated, too—but you probably don't realize that you're just as much a part of the crowding problem as all the other people who decided to come to the beach. (This kind of situation is so common that systems thinkers have identified it as another systems archetype, known as "Tragedy of the Commons.")

In addition to helping us anticipate or even avoid *unintended* consequences, systems thinking can encourage us to reinforce *intended* consequences. For example, an initial success often leads to more success. By investing in early successes (for instance, putting money in a bank account, or spending time with our children), we get dividends later on. Focusing on intended consequences can help us monitor when things go awry. It also helps us develop the patience we need to see our early efforts pay off.

The Systems Thinker's Clock: A Different Look at Time

Systems thinkers view time very differently from how most people do. For one thing, they understand that most living systems don't demonstrate a full "cycle" of their behavior over short time periods. To understand and work with such a system, you have to look at its behavior over a longer time frame. For instance, after years of *perestroika,* the Soviet Union's economic situation changed very little. But as systems thinker Donella Meadows explained, "People are calling it failure, not understanding how long it takes for a nation's capital plant, exhausted soils, and disaffected workforce to revitalize."[4] As another example, it's common for leaders to take credit for any improvements in the economy that occur during their administration. But, those economic changes were likely set in motion way before they took office—perhaps as long as two or three administrations ago.

But thinking long term can involve even longer "time horizons" than just a few changes in leadership. For example, before the Nashua Indians (a native people who lived in New Hampshire) made any decision, they weighed its potential impact on *seven generations* to follow. Or for you gardeners, you know this from the cycles of seed planting, incubation, growth, fruit bearing, eating, and decay.

How can we develop a systems thinker's view of time? We can start by simply being aware of the enormous number of forces in our society and how long they really take to exert their effects. The next time you're reading the newspaper, choose a story and think about how the impact of those events will unfold over the next week, months, years, decades—even generations.

Ferreting out Delays

Another way to learn to take the long view is to practice ferreting out delays. For instance, as frustrated parents of teenagers try to keep in mind, the lessons that you teach your kids about the value of family and community may not "sink in" until the kids reach adulthood.

These kinds of delays (that is, the time lag between your actions and evidence of their effects) can be the biggest reason that unintended consequences happen. Why? Because when we don't see instant results from our actions, we often continue to tinker with the system, coming up with more "fixes" even though we've already taken appropriate action. (We just don't know whether our steps were effective because the results haven't yet made themselves evident.)

For example, if you're like many people, you may battle every day with the hot/cold control in your shower. When the shower water feels too cold, you turn the temperature control to "Warm." But it takes a while (a delay) for the water to warm up. Shivering, you turn the control up even more—to "Hot." But then, when the water finally warms up, it's *too* hot—and you practically scald yourself. So, you turn the control back down to "Cool"—and start the cycle all over again: It takes a while for the water to cool down, so you twist the dial to "Cold." But when you finally get a response, the water's *too* cold. And so on. . . .

Understanding delays can help you avoid this kind of ineffective fiddling with the system. The more you know about delays, the more chance you have of avoiding the kinds of results you had no intention of getting!

Thinking Like a Bathtub

So far, we've been exploring a way to think about systems that focuses on interrelationships, circular causes and outcomes, the impact of delays, and so on. You even saw some simple diagrams that show

how causes and effects can influence each other (for instance, the picture of Jack on p. 27). If you want to know something even more interesting about systems thinking, let's take a closer look at another one of these diagrams.

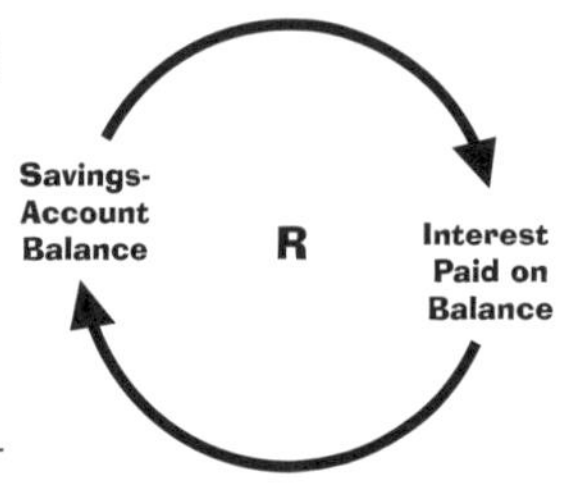

In this diagram, there's something more to the feedback loop showing the relationship between your savings-account balance and the paying of interest on that balance. Let's take a closer look. You can imagine your account balance as a kind of bathtub—the money in it just keeps getting higher and higher (as long as you don't make any withdrawals, of course!). So, the balance is something that *accumulates.* On the other hand, the paying of interest on the account is more like a faucet that flows faster the higher your balance gets. Thus it's like an action, or a process. Systems thinkers would describe your account balance as a *stock* and your interest payments as a *flow.* Each of them influences the other.

Thinking like a bathtub and faucet—that is, thinking of systems as consisting of stocks and flows—is helpful because it lets us understand the more subtle relationships in the system.

Again, systems thinker Donella Meadows offers an intriguing perspective on this. As she explains, we need to understand the difference between the national deficit (a flow—the rate at which we borrow) and the national debt (a stock—the accumulated debt). Reducing the deficit, Meadows points out, will not reduce the level of debt. It will slow down the *rate* at which the debt accumulates—but the debt itself will still keep accumulating.

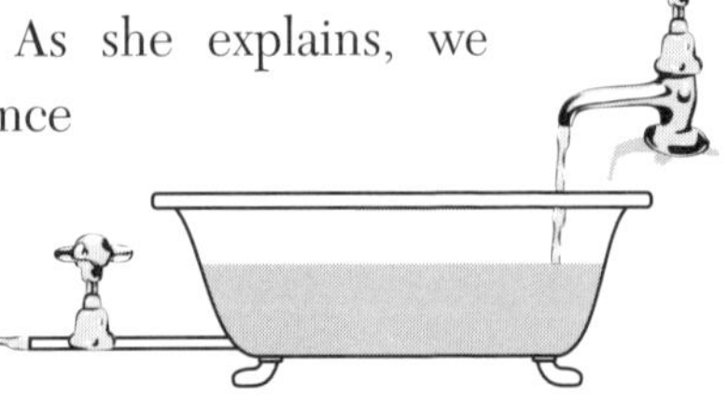

This is a subtle point, but it's important to understand. Most people assume that to increase a stock, you have to increase the flow into that stock (e.g., to accumulate more money in the bank, you have to increase the rate at which you deposit money). But—and here's the surprising part—you can also increase a stock by *reducing the outflow.* As Meadows explains, we can make our nation wealthier by repairing and maintaining old equipment as well as by investing in new equipment.

How to Use This Book

Part II

The stories described in Part 3 of this book are excellent tools for raising kids' awareness of systems' behaviors—in an easy, fun, and educational way. For each story description, you'll find:

Age Range: Each publisher suggests a target age range for their story. But if my own experience using the story tells me that it is also appropriate for readers older or younger than the publisher's target age range, I let you know.

Systems Thinking Concepts: This is a list of the systems thinking–related concepts you might explore through the story. All the concepts are described in Part 1—Systems Thinking: A Means to Understanding Our Complex World (pp. 19–37).

A Quick Look at the Story: This gives a brief summary of the story line: the setting, what happens, who does what to whom, and some additional comments.

Teaching Tips: This section offers suggestions for using the story to teach systems thinking concepts.

Questions to Consider: Here you'll find two sets of questions, one for younger kids (under eight years old) and one for older kids (and adults). Whether you are using these stories with your own children, with students, or even with other adults, these questions should help generate lively conversations.

Voices from the Field: With some of the story descriptions, you'll hear from teachers, parents, and consultants who have used the stories in this collection and have generously agreed to share their experiences.

Partner Stories: These help you reinforce particular concepts and sequence stories in a way that builds on earlier information.

If you're using the books with a group of kids—either at home or in a classroom setting—try having the children read the books in pairs. Or pass one copy of a book around a circle of readers (8–12 maximum), with each person reading one or two pages out loud at a time. If you are working with a large group, try reading the book out loud yourself, showing the illustrations to the group as you go along. Also, educators may find the "Voices from the Field" feature especially helpful. Here are a few more tips for making the best use of this book.

Tip 1: Begin with What Happened

Begin by simply asking readers to describe what happened or is happening in the story. What relationships do they notice in the story, and what happens to those relationships over time? You may also ask readers what they found surprising in the story. (The unexpected often reveals a system at work!) The idea here is to elicit children's perception of the story's dynamics and *then* to get them mulling over the relevant systems thinking concepts.

Tip 2: Trace Cause-and-Effect Relationships

Ask your children or students to draw a causal picture of the story or part of the story, like the one on p. 46. Don't worry about whether they get the loops exactly right. Just encourage them to practice. This activity will get them thinking about circular cause-and-effect loops. After they finish drawing, ask them: What might happen in the story if part A or character B or relationship C were taken away or changed in some way? This question helps kids tune into the systems nature of the story.

Tip 3: Introduce the Language of Causality

In my experience with fifth-grade students, I found that many of them easily adapted to using causal sentence structures. Here are some templates that can help you introduce your kids or students to the language of mutual causality. Try using these as you talk about the stories with young readers:

> If **A** increases, then what does **B** do? Does it increase or decrease? Decrease? Okay, what happens to **C**? It increases? If **C** increases, what happens to **A**?
>
> OR
>
> There were more/less **C** because there were more/less **B**. There were more/less **B** because there were more/less **A**. Because there were more/less **A**, there were more/less **C**.
>
> OR
>
> As **A** got bigger/smaller, **B** got bigger/smaller. As **B** got bigger/smaller, **C** got bigger/smaller. As **C** got bigger/smaller, **A** got even bigger/smaller.

Here's an example taken from ***The Butter Battle Book*** by Dr. Seuss: As the Zooks built more fancy weapons, the Yooks felt more threatened. As the Yooks felt more threatened, they built more fancy weapons. As the Yooks built more fancy weapons, the Zooks felt even more threatened ... and so on (see p. 81).

Tip 4: Ask Causal Questions

As an adult, you can model the kinds of questions that will arouse your children's and students' interest in the systemic nature of a story. For example, try asking: What did the people/animals in the story do? What made them do these things? What happened after they did what they did? How long did it take for those things to happen? If the people/animals do X [cite a possible action], what will happen? What might happen that no one expected? Do some kinds of things keep happening over and over in the story? If so, what? What do you imagine happening to the people/animals in the story 50 years from now? How did you feel when you were reading the story? Did your feelings change as you read? What would you have done differently if you were the main character?

Tip 5: Encourage Kids to Create Their Own Stories

Some kids love to tell stories as well as read them. Ask your children or students to make up some stories and then practice identifying and drawing feedback loops from the stories they create. Use the templates in Tip 3 if appropriate.

Tip 6: Help Children Show What They Already Know

In many ways, children are better and more natural systems thinkers than adults. Try to encourage your kids to reveal what they already know (versus teaching them something new). Help them find meaning in a story and explore how it may relate to their own experiences. For example, encourage kids to give their own personal examples of how something in real life happened to them that was a lot like an event in the story.

Explaining a story's system lessons right off the bat is a sure way to take the enchantment out of the reading experience and deprive kids of their own sense of achievement.

Tip 7: Choosing Stories

In choosing which stories to read, first scan the targeted age range to see which are generally appropriate. Then look at the short list of systems thinking concepts to see which story explores the concepts you or your children have an interest in. You can also use the guide on pp. 10–11 to help you choose which stories to read. Or, read the brief story descriptions to find one that intrigues you or your child. The key thing to remember is that you don't have to use the stories in any particular order. For additional ideas about finding your own systems-oriented stories, see "Finding Your Own Stories" on p. 105.

If You Give a Mouse a Cookie

Author: Laura Joffe Numeroff, illustrated by Felicia Bond
Publisher: HarperCollins, New York, 1985
Format: Picture book, fiction
Age Range: 3–7

Systems Thinking Concepts

Simple interconnectedness, circular feedback, unintended consequences, time horizons, solutions that create new problems

A Quick Look at the Story

This story reveals the unforeseen consequences that can happen when you give a hungry little mouse a cookie. Seems innocent enough. But the next thing you know, the energetic mouse will want a glass of milk. Then he'll want to look in a mirror to make sure he doesn't have a milk mustache. Then he'll ask for a pair of scissors to give himself a trim, and a broom to sweep up. The mouse rascality tumbles on like dominos throughout this delightful book. Unlike dominos, however, by the end of the story we're back to where we started. The mouse wants more milk because he's thirsty after all that exertion, and remembering that a cookie goes well with a glass of milk, he's hungry again.

Teaching Tips

Children delight at seeing the mouse's mischief while also learning the story's moral: You might end up with some big, probably undesirable surprises if things get pushed to an extreme. In real life, many of us tend to ignore unintended consequences, feedback, and delays when making decisions. We all can use practice in paying more attention to these aspects of system behavior. Even though this story is intended for 3 to 7-year-olds, it can also help older children (ages 8–12) practice tracing cause-and-effect relationships to see how an event (giving the mouse a cookie) can feed back on itself. When asked to describe what happens in this story, my nephew (who was 11 at the time), created the picture below.

To explore the systems thinking lessons in this story, first ask kids to think about different kinds of cause-and-effect relationships. For example, there are "domino" relationships (A causes B, which causes C; end of story). You can see dominos in a simple food chain: Big fish eat medium-size fish, medium-size fish eat small fish, and small fish eat insects or plants. (For more discussion of the different shapes of causality, see endnote 2). But there's also the circular kind of causality that we've been discussing (A causes B, which causes C, which comes back and affects A).

Regardless of your readers' ages, try creating some thought experiments that center on feedback. Remind the kids about the nature of feedback—that is, when an action is taken by something or someone, it eventually has some kind of effect on that very thing, person, or group. Then encourage them to imagine some scenarios, such as feeding a pet too many snacks. Ask them what would happen in this particular scenario that might cause them to make a change. Then ask for examples of *desirable* things that feedback can cause. For instance, you gain athletic or musical skill through constant practice (the more you practice, the better you are, the more inclined you'll be to improve, so the more you'll practice, and so on).[5] Or by saving your allowance, you can buy something substantial instead of spending the money on small things as soon as you receive it. Keep in mind that the fundamental causal loop in ***If You Give a Mouse a Cookie*** is the milk and cookie connection (when you give a mouse a cookie, he will eventually become thirsty, which will make him want milk, and want another cookie, which will make him want more milk). The rest of the story can be seen as entertainment or what a systems thinker might call "noise."

Questions to Consider

For Younger Readers

- What happens in the story?
- When have you seen this same kind of thing happen … when one thing makes another thing happen and then another, until you end up back where you started, and you keep going around?

For Older Readers

- Describe or draw a picture of the circular causality in the story.
- What other kinds of situations can you think of where one event eventually feeds back on itself?
- What are the possible unintended consequences of a solution to an everyday problem? For instance, suppose city planners added an extra traffic lane to a crowded highway. Would this produce less traffic or more traffic in the long run? Why?

Voices from the Field

Tim Lucas, director of curriculum and instruction at Glen Rock Public Schools in New Jersey and contributing author to the recently published ***Schools That Learn*** Fieldbook, has used ***If You Give a Mouse a Cookie*** with children in kindergarten through the fourth grade:

"We were able to unearth some important systems thinking language that we could build on, and we introduced causal loops.

One great exercise was to have the students make their own circular story and draw out the major events on a long (four-foot) strip of paper, similar to a cartoon. Then we would make a circle out of the paper by taping the ends together with the pictures inside. If you stuck your head in the paper story and spun the cartoon, you could never tell where the story started or stopped (like a causal loop!).

We hung the paper loops from the ceiling, with some careful stringing at 'kid height.' Students walked around the room, stuck their heads in the paper circles, and spun each others' stories. We were definitely 'spinning tales.'

We also talked about stocks and flows. Students started with the idea of the bathtub, but then began to see other objects in the classroom and school as stocks and identified the flows that impacted them. We started with garbage cans (asking ourselves who filled them and emptied them), lunch boxes, lost-and-found boxes, and even classrooms filled with students. Students figured out that the number of kids in a class (stock) was affected by new families moving in and out of the school community, and house sales in the neighborhood."

Terry Ellen McCarthy, a first-grade teacher in Tacoma, WA, also uses ***If You Give a Mouse a Cookie*** with young readers (what she calls her "little darlin's"). She uses the ***Mouse*** book in conjunction with ***Cinderella,*** and a set of dominos.

"My major curricular focus is to to bring these first graders to literacy in any way I can, since virtually none of them have been read to, or exposed to, literacy in their own homes. A secondary, but equally important personal goal, is to help them understand that there are many life choices outside the ones they see modeled in their neighborhood. They have a right to know, even at this early age, that they can make choices and that those choices will impact their futures for good or for ill. In essence, I have been looking at the potential for systems thinking in the classroom for a while now, without realizing it.

We prefaced our reading of ***If You Give a Mouse a Cookie*** with some brainstorming about the classic story ***Cinderella.*** After brainstorming the sequence of events in ***Cinderella*** and making a simple linear diagram of the story, we set dominoes up in a line, with each domino representing a major 'event' in the story. We toppled the dominoes, and then chatted quite informally about cause and effect. I posed a reflection question to the children about the believability of 'happily-ever-after stories.' I subsequently set the same dominoes up in the same linear fashion, but changed the spacing so that they were closer together.

The children remarked about the reasons why the 'chain reaction' happened more rapidly when the 'events' were in closer proximity. It led to some interesting comments.

The next day, I followed up with ***If You Give a Mouse a Cookie.*** Since this story resulted in a circular feedback loop when we diagrammed it on the board, the kidlettes were able to see that there was potential for this story to revolve in the same circular fashion ad infinitum.

We then had a very informal chat about how we might find that kind of circular cause-and-effect in real life. The children then teamed up and used the dominoes to create their own models of that story."

Partner Stories

Check out these other books by the great author-illustrator team of Numeroff and Bond: ***If You Give a Moose a Muffin*** (HarperCollins Juvenile Books, 1991), ***If You Give a Pig a Pancake*** (HarperCollins Juvenile Books, 1998), and ***If You Take a Mouse to the Movies*** (HarperCollins Juvenile Books, 2000). Another fun book filled with circular stories and poems is ***Arm in Arm: A Collection of Connections, Endless Tales, Reiterations and Other Echolalia,*** by Remy Charlip (Tricycle Press, 1997); picture book, fiction, age range 4–8.

The Old Ladies Who Liked Cats

Author: Carol Greene, pictures by Loretta Krupinksi
Publisher: HarperCollins, New York, 1991[6]
Format: Picture book
Age Range: 9–12 (though this range seems too limited to me; I know readers younger than 9 who enjoy reading this book as well)

Systems Thinking Concepts

Simple interconnectedness, the impact of delays, unintended consequences, goal-seeking behaviors or equilibrium, how a seemingly rational decision can have disastrous large-scale results

A Quick Look at the Story

Inspired by Charles Darwin's observations of the interdependence between clover and cats, this ecological fairy tale shows us, simply, "how things work together." I've never seen a children's book that conveys so well the interconnectedness, mutual need, and delicate balance of the system we call "nature."

The setting is a town, in the middle of an island, that is made up of a unique ecological system: sailors who drink cows' milk, cows who eat sweet red clover, bees who carry pollen to the clover, cats who keep the mice away from the bee honeycombs, and the old ladies who take care of the cats. One day, after the mayor trips over one of the many cats that roam his island home, he decrees that all cats must be kept locked up indoors. But the unexpected happens: Because the cats can no longer keep the mice away from the honeycombs, the chain of ecological interdependence is broken.

Eventually, the sailors who protect the island get sick from lack of milk, and invaders take over. It's up to the old ladies "who know how things work" (what a precious line!) to set things right. Together, they persuade the mayor to rescind the new law. The cats chase the mice into the woods, so the mice no longer eat the bees' honeycomb, so the bees pollinate the clover, so the cows eat the clover and produce milk, so the sailors drink the milk and become strong and healthy—and chase the invaders away.

Teaching Tips

Often when we have a problem, we look for a fast way to fix it (just as the mayor did by banning all cats from the streets at night). While our solution may ease the problem for a little while, it might also have some consequences that we never intended—and that even make the original problem worse eventually. Systems thinkers John Sterman and Daniel Kim have used the well-known story of Helen Keller to illustrate this phenomenon. Helen Keller was blind and deaf as a child but eventually learned to overcome these limitations. Initially, Helen's parents, with the best of intentions, tried to protect their daughter by coming immediately to her aid when she ran into problems. What they didn't understand was that their good-intentioned help was actually having a major unintended, negative consequence: Helen was not learning how to deal with the challenges of day-to-day life.

With the help of her teacher, Ann Sullivan, Helen eventually developed the life skills she needed to care for herself. By moving beyond the "quick fix" of her parent's moment-to-moment interventions to the longer, more fundamental (and more difficult) solution of developing much-needed self-reliance, Helen eventually attended Radcliffe College and became a renowned author and national advocate for differently abled individuals. You can use this story to help kids think about the actions and related unintended consequences they may experience in their own lives.

In addition to providing a way to think about unintended

consequences, this story is also perfectly suited to helping students, young and old, to grasp the idea of dynamic equilibrium. Without having the exact terms for it, we probably are all aware of several "forces of nature" at work on us every day. One of these is dynamic equilibrium; the others are called entropy and negentropy. To talk about these ideas with kids, we could use the phrases "breaking down," "growing," and "balancing."

Entropy is the process by which things break down. These "things" may be the human body, a beloved reading chair, buildings, organizations, etc. When I vacuum the "dust bunnies" under Jack's bed, those dust bunnies (threads, yarn, and other materials from his stuffed animals, blankets, and clothes) are the results of entropy.

Negentropy, on the other hand, consists of those forces that create order and growth. Good examples of negentropy at work are cells (for periods of time), our minds[7], families, and sometimes, economies.

Dynamic equilibrium is another force. In some of the stories, we've seen examples of reinforcing behavior, exponential growth, and up-and-down or back-and-forth patterns (oscillation). A system in a state of dynamic equilibrium maintains its state through a set of counterbalancing forces. Your body temperature is a good example: Your body makes a continual set of adjustments to maintain your temperature at 98.6 degrees. Ideas such as "harmony," "balance," "things are stable," or "if it ain't broke, don't fix it" all hint at systems in a state of dynamic equilibrium.

By encouraging young readers to create a map of the ecosystem described in ***The Old Ladies Who Liked Cats,*** you give them a terrific opportunity to explore the nature of dynamic equilibrium and the impact of human decisions and policies on systems that are in that state.

Questions to Consider

For Younger Readers

- What happened in this story?
- What is connected to what or to whom? (Ask your readers to create a visual map of the whole system, showing how the different characters and animals are linked.)
- When the author says that the old ladies "know how things work together," what do you think she means?

For Older Readers

- What words or phrases might people use when a situation is in equilibrium? Out of equilibrium?
- What would happen if the mayor told people to use pesticides to kill off the bees?
- What "goal-seeking behaviors" do you see in this story?
- Can you think of decisions or policies that were made with good intentions but that have resulted in unintended consequences?

Voices from the Field

Janice Molloy, content director at Pegasus Communications, shares her experience reading ***The Old Ladies Who Liked Cats*** to her young daughter:

"My two-year-old daughter Pamela picked this book off the shelf during her weekly trip to the library. I imagine she was drawn to the playful cats on the cover. As we read the story at home time and again, I wondered how much she was picking up in terms of the chain-reaction of events that led first to crisis and then to resolution for the island's inhabitants. She was interested in the bees and the beehives, and the cats chasing the mice. But, most of all, she was obsessed with why the mayor fell in the mud and why he then made the cats stay in at night!

At first, I thought, 'Well, she's too young to understand the

causal relationships in the story.' But I later realized that she had keyed in on the pivotal moment in the plot: the turning point when the island's gentle balance is thrown into turmoil. The mayor's proclamation that all cats be kept in at night has a series of unintended consequences that affect the quality of life for all. When he finally heeds the advice of the old women 'who know how things work together' and reverses his command, order is eventually restored.

When Pamela is a little older, I'm sure we'll read this book and talk about interconnections and unintended consequences and delays. Maybe we'll even draw a causal loop diagram or two. And I appreciate the message that the old women who like cats are the real heroes in the story. But, for now, we'll continue to talk about the accident-prone mayor and his ill-fated mud bath—over and over and over again!"

Partner Stories

Try contrasting the steady-state nature of ***The Old Ladies Who Liked Cats*** with the reinforcing nature of the loop found in ***If You Give a Mouse a Cookie*** (see p. 45). ***The Lorax*** (see p. 88) also provides an example of an ecosystem knocked off balance by a human policy or decision.

For another book based on a true story of ecological disruption, see ***The Day They Parachuted Cats on Borneo: A Drama of Ecology*** by Charlotte Pomerantz (Young Scott Books/Addison-Wesley, Massachusetts, 1971, ages 8–12). Written in poetry, this book tells how spraying for mosquitoes in Borneo affected the entire ecological system—including cockroaches, rats, cats, geckoes, the river, and eventually the farmers. There is also a short story on the topic for older readers called "Top of the Food Chain" in ***Without a Hero (and Other Stories)*** by T. Coraghessan Boyle (Viking, Penguin Books, New York, 1994).

The Cat in the Hat Comes Back

Author: Dr. Seuss
Publisher: Random House, New York, 1958
Format: Picture book
Age Range: 4–8 (but open-ended possibilities for grown-ups, too!)

Systems Thinking Concepts

Simple interconnectedness, unintended consequences, the archetypal systems story of "Fixes That Fail"

A Quick Look at the Story

In this tale, the mischievous cat from ***The Cat in the Hat*** is back! He returns on a snowy day when Sally and her brother have household chores to do—especially shoveling snow. But as we know from ***The Cat in the Hat,*** this cat has other ideas of fun in mind—in this case, to eat cake in the children's bathtub. Seems harmless enough. "'I like to eat cake / in a tub,' laughed the cat / 'You should try it sometime,' / Laughed the cat as he sat."

But there's a problem: The young boy sees that the cat has left "a big long pink cat ring" in the tub. To clean it up, the cat uses the children's mother's *white dress*—promptly transferring the stain to the dress! Fearing that the spot might never come off the fabric, the brother and sister urge the cat to do something. So the cat flings the dress against the wall. The pink stain is gone from the dress … but now it's on the wall! "Now the dress was all clean. / But the wall! What a mess."

Then the spot ends up on Dad's shoes, then on the rug, then on the bed, until finally, even the cat doesn't know what to do. He

calls in reinforcements. From inside his hat come Little Cat A, then Littler Cats B, C, D, E, and so on. The army of cats manages to clean up the house, but only by covering the snow outside with thousands of pink spots. Out from the hat come more tiny cats—Little Cats N through V—who by shooting the spots with pop guns only make matters worse.

Just when we think Sally and her brother will never extract themselves from this "wicked mess," Little Cat Z saves the day with a special "Voom" machine: The front walk gets shoveled, and Sally and her brother are left to wonder just when that rascal cat will show up again.

Teaching Tips

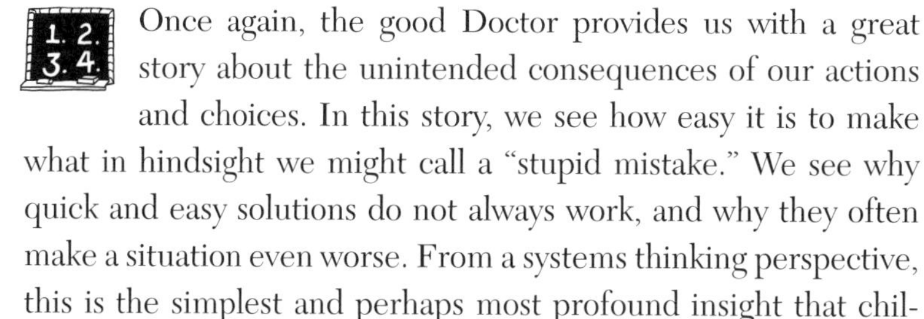

Once again, the good Doctor provides us with a great story about the unintended consequences of our actions and choices. In this story, we see how easy it is to make what in hindsight we might call a "stupid mistake." We see why quick and easy solutions do not always work, and why they often make a situation even worse. From a systems thinking perspective, this is the simplest and perhaps most profound insight that children can learn.

In reading ***The Cat in the Hat Comes Back,*** children laugh along as they watch the cat, with what we assume to be the best intentions, try to solve the problem of the moving spot. But what happens? By shifting the problem around, the cat only makes matters worse. If you've ever lived in a small apartment with limited closet space, you might recognize this situation: You move a pile of clothes from one chair to the next, never actually finding a place to put the clothes away. In the process of moving, the clothes become disheveled and make the small space appear even more cramped!

The systems thinking archetype that comes to mind in this scenario is "Fixes That Fail." In this archetype, a problem symptom (in this case, the tenacious pink spot) is addressed by a "fix" (moving it to another location). As the quick fixes continue (including shooting the spot with pop guns), a set of unintended consequences occurs: The pink spots only proliferate, covering all of the outdoors.

As systems thinker Daniel Kim says, when you're caught in this kind of dynamic, "there is a feeling that you need to try the same solution just a little more, and then a little more, and then one more time ... until you catch yourself resisting the idea of trying anything else." A sense of powerlessness can also characterize this kind of situation: It's clear that your solution isn't working, but there seems to be no other choice for addressing the problem. Borrowing money when you find yourself in debt is another example: You still need more money to pay the usual bills.

The cat's and the children's mental models, or assumptions about how the world works, offer another good thread to follow in this story. When we are trying to help children think about how to solve real-life problems, we want them to make a subtle shift and realize that how they *think* about a problem makes a difference in how they try to solve it. In this story, Sally, her brother, and the cats are all mired in the same powerful mental model: The solution to the problem is to move the pink stain "out of sight, out of mind." Yet every attempt to do that only makes things worse.

In situations like this, systems thinkers like to look for what they call "leverage points"—actions that can create positive change. The difficulty is that our perceptions of a problem often obscure the leverage points. We can think of Little Cat Z's "Voom" machine in this way. As environmentalist Don Robadue, who has used ***The Cat in the Hat Comes Back*** with other adults, explains:

"The Voom machine is a symbol of the leverage points in a dynamic situation. Sometimes it's better not to flail around doing whatever comes to mind. Instead, we need to find the usually very small, often hidden factors that lie behind symptoms. We are too often held captive by our rigid mental models of what the right answer is supposed to be, and thus we inadvertently keep our problem stuck firmly in place."

Questions to Consider

For Younger Readers

- What happens when the cat tries to solve the "pink spot" problem?
- Can you think of a time when someone was trying to be helpful but only made matters worse?

For Older Readers

- How are the cats and the children defining "the problem"? How might this definition actually get in their way of solving the problem?
- What do Sally and her brother do as the spot problem gets worse?
- How would you describe the children's and the cat's problem-solving approach? What impact do their various solutions have on the original problem? (You might point out here that as the problem gets worse, the characters call out reinforcements. But these stepped-up approaches to solving the problem only make the problem worse.)
- Can you think of a real-life example of this kind of problem-solution escalation?

Voices from the Field

Don Robadue is an environmental planner and trainer at the University of Rhode Island's Coastal Resources Center. He has used Dr. Seuss stories with environmental professionals around the world to help them think through the dynamic political and social situations surrounding the search for a balanced use of the world's coastal ecosystems. Here, Robadue describes his experience using ***The Cat in the Hat Comes Back:***

"Every other year, the Center offers a four-week, international training program for scientists and professionals who are interested in how to design programs to manage coastal resources such as

beaches, bays, coral reefs, and wetlands. I wanted to introduce systems thinking ideas throughout the course. I chose to do this through storytelling, focusing on the tales of Dr. Seuss, who is well known not just in America but throughout the world.

The great thing about a story like ***The Cat in the Hat Comes Back*** is that it provides both entertainment and revealed wisdom. The narration conveys an approach to problem solving that we can add to our intuition quickly because the story setting is literally unfamiliar yet psychologically authentic. This intentional disorientation is a key to letting us shake free of our current mental models so that we can see more clearly how things really are. Professionals in my field must be able to work with all kinds of people with different interests. They must help them get beyond their own narrow interests and move toward solutions that address root concerns in an effective manner. This is best done through a process of joint inquiry, learning, and experimentation."

Partner Stories

For a simple story of escalating, unintended consequences, see ***Why Mosquitoes Buzz in People's Ears: A West African Tale*** by Verna Aarderna, illustrated by Leo and Diane Dillion. This breathtakingly beautiful book (and Caldecott award-winner) is geared toward 4 to 8-year-old readers (Dial Books for Young Readers, 1975).

Once a Mouse: A Fable Cut in Wood

Author: Marcia Brown
Publisher: Charles Scribner's Sons, New York, 1961
Format: Picture book, fiction
Age Range: All

Systems Thinking Concepts

Simple interconnectedness, unintended consequences, multiple causes and effects

A Quick Look at the Story

In ancient India, a noble prince, or "rajah," is said to have collected popular animal fables as a way to instruct his errant sons. In this fable from the *Hitopadesa* (a set of classic Indian fairy tales and fables written between A.D. 400 and A.D. 1100), a hermit sits reflecting on the concepts of "big and little." He sees a mouse about to be snatched up by a crow, and decides to help the little animal by scaring away the crow. But soon the mouse is threatened by a cat, so the hermit, "mighty with magic," turns the mouse into a bigger cat. Eventually, the mouse becomes increasingly vain as the hermit helps it change from mouse, to cat, to dog, to tiger. The proud tiger becomes, as a Westerner might say, "too big for his britches," terrorizing the forest and threatening the hermit. The hermit, realizing that the tiger (now blinded by his own size) has lost his sense of connection to the other animals and his place in the forest family, finally turns him back into a mouse.

Teaching Tips

This tale about basic morality was most likely written to encourage readers to consider the dangers of unbounded pride and the wisdom of humility. For aspiring systems thinkers, it also provides a circle of causes and effects, suitable for tracing by young readers. From a philosophical perspective, it calls into question the widespread (at least in the West) notion that "bigger is better." It also resonates with the idea that in social systems, as in any living systems, nothing grows forever. Seemingly uncontrollable growth will always eventually run into some sort of limit—and a balancing process will kick in to bring things back under control.

To get readers thinking about this, try talking with them about common predator/prey relationships; for example, deer and wolves. If the winter is mild, the deer population might increase rapidly. But if this happens, the wolves find it easier to catch food. So, more and more wolves remain where there are lots of deer. When the wolf population gets large enough, the wolves begin catching more deer and reducing the deer population. The cycle continues: When there are fewer deer, the wolves can't catch and eat as much food, and eventually their numbers decrease. Then, over time, the deer population increases once again. And so on. . . .[8]

In ***Once a Mouse,*** the limit or constraint on the mouse's seemingly infinite growth is set by the humble hermit. In real life, the goal is to plan for and manage limits ourselves, or as Russell Ackoff says, "plan or be planned for." Otherwise, nature or the system itself will step in and do it for us—often in very unpleasant ways. (For example, in a number of rapidly growing suburban communities, neighborhoods have expanded so much into wild animals' habitats that coyotes and other animals are being forced to feed on the trash that people leave outside. Sometimes they even grab pets that people allow to wander outside the house.)

Questions to Consider

For Younger Readers

- What happens in the story?
- When the hermit was sitting and thinking about "big and little," what do you think was on his mind?
- Can you think of other situations when somebody *meant* to do something good, but they caused something bad to happen instead?

For Older Readers

- Are there times where bigger or more is not always better?
- Can you think of other situations when something kept growing and growing—to the point where things got out of control?
- If yes, what was the limit that stopped the growth, and how was it imposed?

Partner Stories

To help kids learn more about the problems of uncontrolled growth, follow ***Once a Mouse*** with a reading of ***A River Ran Wild*** (see p. 77). This book, about the eventual overuse of the Nashua River, provides a real-life example of what can happen when growth runs amok. Also, ***The Lorax*** (see p. 88) is a fantastical story that points to the pitfalls of the particularly Western notion that "bigger is always better."

The Sneetches and Other Stories

Author: Dr. Seuss
Publisher: Random House, New York, 1961
Format: Picture book, fiction
Age Range: 4–8, but compelling to readers of all ages

Systems Thinking Concepts

Simple interconnectedness, unintended consequences, balancing feedback loops, oscillations, the way that a system's structure drives events, seeing how a quick fix can cause a side-effect that makes the problem worse (the "Shifting the Burden" systems archetype)

A Quick Look at the Story

This gem from Dr. Seuss shows how prejudice and the drive for exclusivity can result in wasted energy and depleted resources. Star-Belly Sneetches are fuzzy green animals with neon-green stars in the middle of their stomachs. Plain-Belly Sneetches have no star. Just as bell-bottoms, mini-skirts, Izod shirts, and a Tommy Hilfiger label have (at various times) made some students feel superior, so the small, green star lets some Sneetches brag, "We're the best kind of Sneetch on the beaches."

Eventually, an enterprising imp ("Sylvester McMonkey McBean") cashes in on the situation. For a pretty penny, he adds stars to the Plain-Belly Sneetches with his peculiar machine. Suddenly green stars are everywhere! To remain distinctive, the Star-Belly Sneetches go through the imp's "Star-Off Machine." The cycle continues until Sneetches of all kinds have spent every

last cent of their money to one-up each other. Finally, the Sneetches learn to accept their differences and themselves and so put McBean out of business.

Teaching Tips

This story can be used to practice "lowering the water line" and to introduce another interesting systems archetype called "Shifting the Burden." Here's how the archetype played out for the Sneetches: Rather than solving their problem by learning to accept their differences (perhaps the best—but also the most difficult—solution), the Sneetches tattooed and untattooed stars on their bellies (an easier but ultimately expensive "quick fix"). In the drawing below, you can see how this situation would look in a causal loop diagram.

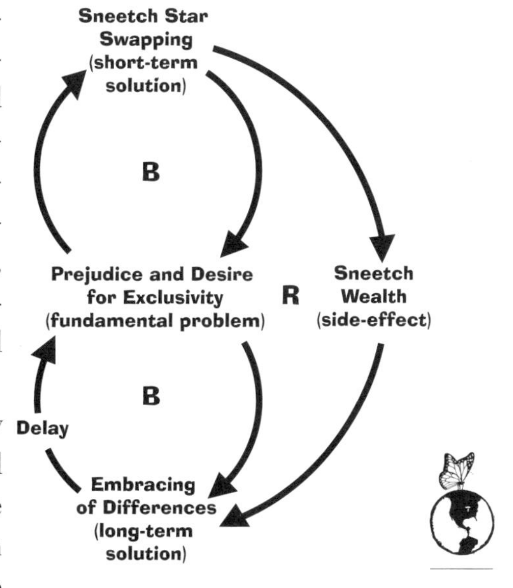

A "Shifting the Burden" situation usually begins with a problem. In this case, the tattooed stars have become a status symbol; that is, if a supposedly low-class group has them, the high-class group *doesn't* want them, and vice versa. (This, unfortunately, is an unattractive but real aspect of human nature.)

In this story, the Plain-Belly Sneetches want to be "in," and the Star-Belly Sneetches (the "in" group) want to keep them "out." This situation prompts the Plain-Belly Sneetches to solve their problem by tattooing themselves with stars. This "quick fix" relieves the Plain-Belly Sneetches' problem symptom in the short run, but it eventually traps all the Sneetches in the "in/out" cycle. It also diverts the Sneetches' attention away from one *real* source of their problem, which is their inability to accept and embrace their differences.

In addition, the "quick fix" creates a troubling reinforcing cycle: The Sneetches run out of money ("Sneetch Wealth decreases"), as they pay the owner of the Star On/Off machine more and more to tattoo and untattoo themselves. In real life, these sorts of side-effects often make it even *more* difficult to put a fundamental, longer-term solution into action.[9]

This story also has a lot to teach about the nature of simple balancing loops. Because balancing feedback loops seek stability, they often create oscillations—some big and some small—around some goal. (Remember the example of body temperature? Your body temperature might get a little higher or lower than 98.6 degrees—but your temperature-control system almost always brings it back to normal.) The problem is that those who get caught up in this sort of situation (like the in/out cycle experienced by the Sneetches) often can't see the cyclical dynamics they are experiencing—and causing. This is particularly true if the cycles stretch out over several years. For example, today my doctor friends tell me there are more than enough doctors in the United States. As people talk about this big surplus, more young people steer clear of the medical field. Within a few years, however, this creates a shortage. Jobs go begging, and young people are once again urged into the field—which then creates another round of surplus. Yet most of us have short memories and tend to think that whatever is going on now will continue indefinitely into the future.

Questions to Consider

For Younger Readers

- What is happening here?
- Have you seen this kind of thing before? Can you make up a story that sounds similar to this one?
- If you could visit the Sneetches, how might you try to help them with their problem?

For Older Readers

- What happened in the story?
- Think about Sylvester McMonkey McBean, the owner of the Star On/Off machine. By exploiting the fears of both groups of Sneetches, he drains their bank accounts. Can you think of other real-life people or organizations that have this ability? Provide some examples.
- One reinforcing element in this kind of situation is peer pressure. If one member of a group decides to get the "star," all the others follow along. What can be done to reverse or balance such spirals?
- Who benefits most from the "in/out" competition that the Sneetches have going on?
- What do you think the two groups of Sneetches were thinking when they ____________? (Fill in the blank by citing different points in the story.)
- What else could make it hard for the Sneetches to see what's going on?
- What else (besides the Sneetches' losing all their money) could make it *easier* for them to see what's going on?

Partner Stories

The Butter Battle Book by Dr. Seuss (see p. 81) and ***The Giving Tree*** by Shel Silverstein (Harper & Row, Publishers, 1964) can also be viewed through the lens of "Shifting the Burden." (In the latter story, the boy grows into an adult, and shifts the burden of finding his own happiness onto an aging, resource-depleted tree.)

Anno's Magic Seeds

Author: Mitsumasa Anno
Publisher: Philomel, New York, 1992
Format: Picture book, fiction
Age Range: 6–10 (Note: The story asks readers to do a series of mathematical computations as they follow along. Younger readers—ages 5 and under—will enjoy it, but most will not yet grasp the math behind the story.)

Systems Thinking Concepts

Feedback loops, unanticipated limits to growth, delays, boom-and-bust cycles, conservation of resources, exponential growth

A Quick Look at the Story

A wizard gives a happy-go-lucky farmer named Jack two mysterious golden seeds. He instructs Jack to eat one, which will sustain him for a full year, and to plant the other. Jack obeys, and the plant grows, eventually bearing two seeds. The following year, Jack plants both new seeds—and the new plant produces *four* seeds. He eats one seed and plants the other three—and reaps *six* seeds the following year. As the years go by, Jack marries, raises a family, and continues to plant all but one seed from every crop. His crop of seeds doubles every year. After a terrible storm, Jack and his family emerge from their house to find that the storm has washed the crops and storehouses away. Happily, Jack has saved enough seeds to feed his family and start planting once again.

Teaching Tips

This Japanese folk tale is fundamentally a celebration of the mystery and magic of seeds, plants, and nature itself. But it can also be a wonderful way to explore an often subtle but key behavior of systems: rapid or "exponential" growth or decline.

In a note to his readers, Mitsumasa Anno explains: "I have called this book ***The Magic Seeds*** because, in fact, there is a mysterious power in even one tiny seed that seems quite overwhelming." Albert Einstein would agree. When asked to name the greatest force in nature, Einstein replied: "compound interest." Like the seed in this story, a dollar in a bank account has the same potential for steady doubling over time.

But as boom-and-bust cycles in the economy have shown us, nothing grows forever. In this story, a limit is introduced—the storm—which cuts off the tremendous growth of Jack's crop.

Why is it important to be aware of exponential growth processes? The reason is that most of us significantly underestimate exponential growth. The ancient Chinese philosopher Han Fei-Tzu (ca. 500 B.C.) was well aware of this bottleneck in our perceptions when he spoke about hidden consequences of population growth:

> ***"People at present think that five sons are not too many and each son has five sons also, and before the death of the grandfather there are already 25 descendants. Therefore people are more and wealth is less; they work hard and receive little."***

Han Fei-Tzu pointed to a quality of exponential growth loops that can be tricky to grasp: The larger the quantity (e.g., number of seeds, amount of money in a bank account, number of children in a family), the greater the rate of growth—and, owing to the reinforcing process, the greater the quantity.

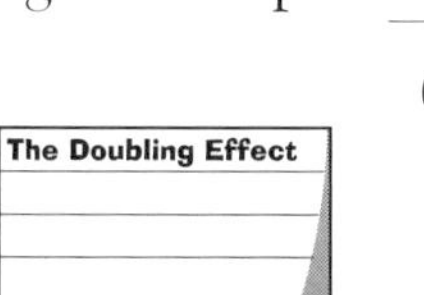

One reason this type of growth can be hard

to detect is that it takes time for what is called the "doubling effect" to show up.

This simple French riddle helps us grasp the idea of the doubling effect:

Suppose you have a pond in your backyard on which a water lily is growing. The lily plant doubles in size each day. If the lily were allowed to grow unchecked, it would completely cover the pond in 30 days, choking off the other forms of life in the water.

*For a long time, the lily plant seems small, and so you decide not to worry about cutting it back until it covers half the pond. On what day will that be?**

Questions to Consider

For Younger Readers

- What happened in this story?
- What might have happened with Jack's seeds if the storm had never come?
- Where else have you seen this happen—when something grows so much and so fast that it seems it will never stop?

For Older Readers

Ask the questions above, as well as the following:

- What might have happened if Jack had followed the wizard's instructions, and planted only one seed?
- What signs in the story hint that exponential growth is happening?
- What else do you see growing exponentially or decaying exponentially in the world around you? (Kids' answers may include: rumors, disease, populations, fads.)

Partner Story

Try contrasting ***Anno's Magic Seeds,*** as a story of limited exponential growth, with ***Tuck Everlasting*** by Natalie Babbit (Farrar, Straus and Giroux, 1975), a chapter book for ages 9–12. In this exciting adventure, a 10-year-old girl discovers a magic spring, which turns out to be a fountain of youth. The story explores the lure and implications of the ultimate example of unlimited growth—the ability to live forever. For another mathematical folktale about exponential growth, try reading ***One Grain of Rice*** by author-illustrator Demi (1997, Scholastic Press: New York). This picture book is suitable for ages 5–9, but enjoyable for younger children if read aloud.

**The answer to the water-lily riddle? On the 29th day. You have one day to save your pond.*

Zoom

Author: Istvan Banyai
Publisher: Viking Children's Books, New York, 1995
Format: Picture book, nonfiction
Age Range: All

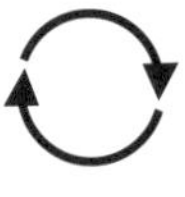

Systems Thinking Concepts

Understanding the impact of our own perceptual filters, multiple levels of perspective, "nested systems," systems boundaries

A Quick Look at the Story

The first thing you realize when you open ***Zoom*** is that there is no "story" per se. Instead, an imaginary camera pulls you back from scene after scene, and you find your perspective changing from page to page.

In this wordless book, we first see a close-up shot of a rooster's comb. Zoom out a bit more, and you see two children watching the rooster. Zoom out again, and you see that the two children are on a farm. Then, you see a large hand appear. You realize that this isn't a real farm, but a play farm set. You keep zooming out, and you realize that the girl playing with the farm set is really a picture on the cover of a magazine, which is being held by a boy, who's sleeping in a chair, and so on—until finally you see the earth growing smaller and smaller. If you've seen the video ***Powers of 10*** (see Additional Resources, p. 115), you've experienced this feeling before—that of looking at a painting and then finding yourself *in* the painting, then being in a photograph of yourself looking at the painting, etc.

Teaching Tips

Zoom takes you on a mind-bending visual journey. But how can it help children to think about and even better see the systems around them? At its most fundamental, ***Zoom*** makes us examine our frames of reference and ask, "Am I sure about what I thought I saw?" As we turn each page, we realize, in real time, the limiting or incomplete assumptions we made about the previous picture. As this perspective-altering process unfolds, we become more and more aware of our assumptions—and how wrong they can be.

The more we can *have* (or see) the frames we are using, rather than be *had* by them, the more we can check how well they match reality. As Ernst Bombrich, author of ***The Story of Art*,** has said:

> ***"Seeing depends on knowledge***
> ***And knowledge, of course, on your college***
> ***But when you are erudite and wise***
> ***What matters is, to use your eyes."***

Zoom is an excellent book for realizing that where we *stand* in a system affects what we *see*. Why is this awareness important for systems thinkers? It can help us look at a system from various levels of perspectives—e.g., a student's perspective, an administrator's, a teacher's, a parent's, a principal's, and so on—and better see how a system's behaviors arise from interactions among all those levels. For example, a teacher's incentive (provided by an anxious school administration) to improve students' performance on high-stakes tests may be at odds with a parent's efforts to keep their child's homework load in check.

Zoom can also teach us to think about system boundaries. The skin is a good example of a boundary for the human body. When we want to understand a system, we need to define it by clarifying what we see as its boundary. By consciously choosing a boundary, we can focus on the factors and inter-relationships that most directly influence the behavior of interest.

If we define too large a boundary, we might overwhelm ourselves or miss the story we are most interested in. For example, if we want to understand why the student drop-out rate is increasing in one school, gathering data about the administration's and teachers' economic, social, religious, and political backgrounds is most likely too wide a frame and misses a key element in the story: the experience of the student. At the same time, if we define too small a boundary in trying to understand a system, then we may not have *enough* information or see key inter-relationships.

A nice illustration of drawing boundaries can be found in a satellite-based research experiment conducted some years ago by NASA. In 1978, the satellite Nimbus 7 was launched into the stratosphere to gather long-term data on significant changes in the atmosphere high above the earth. However, when they designed the experiment, they assumed that ozone concentrations were unimportant because they could not change. The boundary was set too narrowly. Consequently, the computers on board were programmed to suppress any information about the variation in ozone levels that the satellite did sense. Therefore, data were not transmitted back to earth that would have given us several years' earlier warning about the great damage being done to the ozone layer by chlorinated hydrocarbon chemicals.[10]

Questions to Consider

A simple way to demonstrate the idea of frames is to ask kids (or grown-ups) to create a small circle by touching their forefinger to their thumb.

Then ask them to look through their "frame," first holding it at arm's length from their face, and to focus on a specific object (e.g., a picture on the wall or a person sitting across the room). Ask some of the questions described on p. 75. Then, keeping the same object centered within the hole, ask them to bring the frame halfway to the eye, and then close to the eye. Ask some of the same questions at each stage.[11]

For Younger Readers

- What do you notice when your finger frame is close to you?
- How about when it is far away?
- Do you notice different things?

For Older Readers

- What do you notice? (Ask this question at every stage.)
- What questions could you answer with the information available to you through your frame?
- Who might be interested in the data you are gathering? For example, what professions might be interested in close-up data, versus gathering information from a broader perspective?
- What would be an action you could take to change what you see?
- Have you ever thought you were seeing something accurately but realized that, with a different perspective, you would see it differently?
- What helps you to alter your perspective?

Voices from the Field

Tracy Benson, a regional coordinator with the Waters Foundation's Systems Thinking and Dynamic Modeling K–12 Education project, has used this book as a way to talk about expanding spatial and temporal boundaries:

"I used ***Zoom*** in a workshop of school administrators. We were looking at student achievement, using test scores as the primary indicator. I wanted to help them expand the boundaries of their thinking. I thought ***Zoom*** would help them try to expand their view both spatially and temporally beyond the scope of one annual set of test scores (the drops and increases) before they made major decisions about next steps. We had behavior-over-time graphs of trends in test scores (on overhead slides) and we placed various 'mat-type' versions of frames (various sized openings) on the overhead over the graphs. The administrators realized that they were focused on one-year increases and drops in test

scores and were not paying attention to the wider view of trends (i.e. change over time). The process helped them to slow down and examine the system with a wider lens before making significant decisions involving curriculum, instruction, and budget."

Partner Stories

Check out ***ReZoom,*** also by Istvan Banyai (Viking Books, 1998), a companion to the original ***Zoom*** book where again nothing is as it seems. The 10-minute video ***Powers of 10*** is another excellent way to explore the notion of "levels of perspective" and the idea of systems nested within other systems. Starting at a picnic by the lake in Chicago, the film travels, by powers of 10, to the outer edges of the universe. Every 10 seconds, we view the starting point from 10 times farther out. When we see the galaxy as a speck of dust, we're then taken back to earth, and into the hand of the picnicker, again by powers of 10, until finally the journey ends inside the proton of a carbon atom within the picnicker's white blood cell. For an experiential exercise related to these ideas, see the "Circles in the Air" exercise in Volume I of the ***Systems Thinking Playbook*** (Turning Point, 1995/1996).

A River Ran Wild

Author: Lynne Cherry
Publisher: Harcourt Brace Jovanovich Publishers, Orlando, Florida, 1992
Format: Picture book, nonfiction
Age Range: 6–10

Systems Thinking Concepts

Simple interconnectedness, feedback, reinforcing and balancing loops, balancing the short term with the long term, "Limits to Success," systems nested within other systems

A Quick Look at the Story

This is the true story of the life and near death of the Nashua River, from its valley's settlement 7,000 years ago by a tribe of Algonquin Indians until its relatively recent reclamation and revitalization. This beautifully illustrated book takes us from the discovery of the river to the arrival of English settlers, industrialization and the dumping of dyes into the river, the river's near devastation, and finally to a successful effort, begun in the 1960s, to clean up the river and restore its wildlife. With its border of pictures of significant events, people, and animals on key pages, the book also does a wonderful job of illustrating the many interconnected ecosystems—which include creatures from fish to moose to bald eagles—and social systems that live in and around the river.

Teaching Tips

This story is a treasure trove of systems lessons. At its most fundamental, it is the compelling tale of the inter-relationships (which the English settlers had taken for granted) among the Nashua people, the settlers, and the environment and inhabitants of the New Hampshire and Massachusetts woodlands and riverways. As with ***Once a Mouse,*** a desire for unlimited growth leads the protagonists to create a web of unintended consequences. ***A River Ran Wild*** describes a compelling, real-life example of the systems principle that "nothing grows forever."

The figure below shows what the "Limits to Success" archetype looks like in general (a reinforcing loop coupled with a balancing loop).

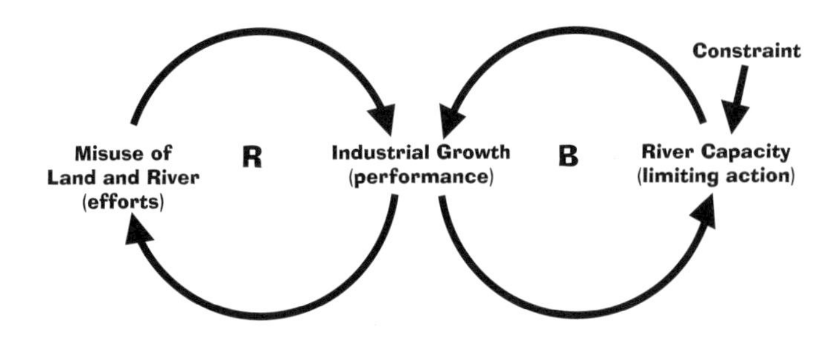

In this story, the source of rapid growth is the English settlers' desire to inhabit more and more of the land surrounding the Nashua River. But because nothing grows forever, the settlers eventually run into a limit—the Nashua River's capacity to be used fairly and effectively (as shown in the balancing loop in the above drawing).

Through this story, you can encourage your child to think of herself as a "third-party helper." She can imagine stepping into the story as a wise traveler who comes upon the Nashua Indians and newly arrived settlers and sits down with both groups. By helping them to map out the causes behind the growth and the potential danger points, she can also help them to anticipate future problems and eliminate them *before* they become threats.

The wise helper might talk with the two groups about time horizons—that is, encourage them to think beyond this day or the next. For instance, how might the settlers' and Indians' views change if they thought about the Nashua River's situation 50 years in the past, 50 years into the future, and in the present—all simultaneously?

How might your wise helper also persuade the two groups to *change* their behavior? Instead of punishing undesirable behavior, she might tempt them with desirable *visions* of what could be. In this story, environmental activists did stage many protests, but they also painted a compelling picture of what the river *could* be like ("a sparkling river full of fish" versus a "stinking, smelly sewer"). How? By showing politicians jars of dirty water and persuading the paper mills to build waste-processing plants.

Questions to Consider

For Younger Readers

- What happens in this story?
- How did the river become polluted?

For Older Readers

- How many years go by during the story?
- Are the Nashua Indians focused on the past, the present, or the future—or all three? What about the English settlers? Are both groups thinking about time in the same way?
- What kinds of problems happen as the settlers use more and more of the land and the river?
- What finally stops the people from using more of the land and river?
- What would have happened if the misuse of the river had continued?

Partner Stories

For other stories about the interaction of people and nature, see ***Who Speaks for Wolf*** by Paula Underwood (see p. 93) and ***The Lorax*** by Dr. Seuss (see p. 88). ***Brother Eagle, Sister Sky*** by Chief Seattle (illustrated by Susan Jeffers, Dial

Books, 1991) is a children's version of Chief Seattle's famous speech in which he describes his people's love and respect for the earth.

The Butter Battle Book

Author: Dr. Seuss
Publisher: Random House, New York, 1984
Format: Picture book, fiction
Age Range: 8 and up, but excellent for adults, too

Systems Thinking Concepts

Simple interconnectedness, circular feedback, reinforcing loops, "Escalation," the impact of the interconnections among a system's parts (structure) on the everyday patterns and events we see happening around us, the power of small decisions to lead to big problems

A Quick Look at the Story

In this story, Dr. Seuss tells of a feud between the Yooks and the Zooks. What's the source of the conflict? One clan eats their toast butter-side-up; the other eats it butter-side-down! Members from each group build a wall to keep the two clans apart and begin feuding with hand-held slingshots. Eventually, they move on to more sophisticated weaponry, until each side has the capacity to destroy each other, and the world!

Teaching Tips

This story sets up a compelling analogy for any situation such as the Cold War arms race—where tensions and even violence just keep getting worse. From a systemic perspective, the intensifying feud between the Yooks and Zooks offers powerful lessons about escalation and possible ways to stop cycles of violence and aggression.

The "Escalation" archetype generates some of the most troubling problems facing us in society. It's also one of the more commonly occurring systems archetypes. Young children immediately understand this archetype if you give them the example of two bullies fighting on a playground. One shoves the other, and the other shoves back harder, until an all-out brawl ensues. In organizational life, we adults might recognize the story of escalation in typical price wars. Or in a more deadly confrontation, the "Escalation" structure can lead to catastrophic consequences such as those faced by U.S. president Kennedy and Soviet chairman Khrushchev during the Cuban Missile Crisis.

You can see the reinforcing nature of the conflict in this story

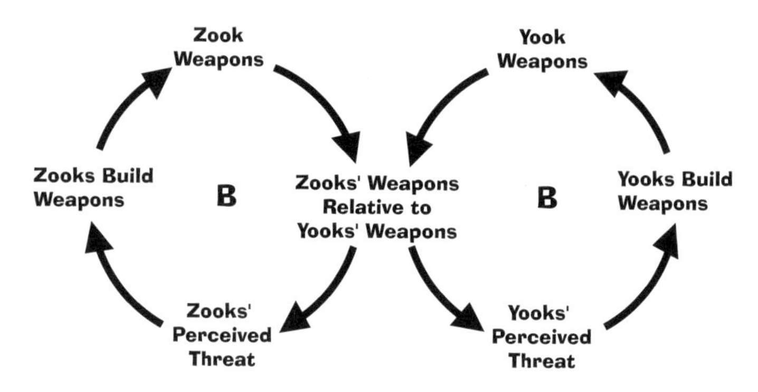

in the two intersecting loops illustrated below.

The conflict grows as one party (the Zooks) takes actions (builds a weapon) that the other perceives as a threat. ("A very rude Zook by the name of VanItch snuck up and sling-shotted by Snick-Berry Switch.") As the other party's (Yooks') sense of security and superiority diminish, they feel threatened and respond in a similar manner, building a Triple-Sling-Jigger and increasing the threat to Zooks. As a result of this move, the Zooks feel threatened and take even more intimidating actions. You can see the reinforcing nature of the conflict by following the outline of the figure-8 produced by the two balancing loops in the above figure.

Questions to Consider

For Younger Readers

- What do you think is happening in this story?
- What do you think the Zooks and Yooks were thinking when they ________? (Fill in the blank by citing different points in the story.)
- What could the Zooks and Yooks do to get along better?

For Older Readers

- What happened? What do you think might happen next?
- What other stories, either from real life or from fiction books, describe the same kind of behavior?
- Who or what do you think is being threatened here? What is the source of the threat?
- What could the characters do to resolve their feud?
- If you were an international peace negotiator assigned to help these two tribes avoid mutual destruction, what would you suggest the Yooks and Zooks do to counteract the escalating pile of weapons? Or even better, what could they do to *reverse* the effects of the escalation? For example, what role might the people in the "backroom" play in your plan?

Voices from the Field

I've read this story with a number of groups, including fifth-grade students, educators in the western U.S., and system dynamics researchers and consultants in Europe. The fifth-grade students (who greeted me as the "loopy lady") were in the midst of learning about fractions, percents, and decimals. Graphing had already been covered in the fourth grade. In English class, they were reading a series of Native American folktales. They were familiar with the idea of ecosystems, but the official unit on that topic was to come in the sixth grade. In computer class, they were being introduced to word processing and effective use of the

Internet, and in social studies, they were investigating the Europeans' arrival in America and its effect on Native Americans. Just prior to one of my class visits, the teacher had framed the bombing of Iraq by the U.S. as a "moral dilemma" and, I was told, an all-day debate ensued. When I sat down to read ***The Butter Battle Book*** with small groups of these students, I found, to my own hidden delight, that a number of children made astute connections between the book and to the stories of war that were making newspaper headlines at the time. For example, one student began to trace an imaginary spiral with his finger, saying, "They do something worse to each other each time, over and over."

Partner Story

For another story that addresses escalating behavior, see ***Billibonk and the Thorn Patch*** by Philip Ramsey (Pegasus Communications, 1997).

Tree of Life: The World of the African Baobab

Author:	Barbara Bash
Publisher:	Sierra Club Books and Little, Brown and Company, New York, 1982
Format:	Picture book, nonfiction
Age Range:	4–8, but a delight for people of all ages

Systems Thinking Concepts

Simple interconnectedness, awareness of time horizons, circular feedback loops, delays

A Quick Look at the Story

This beautiful picture book documents the life cycle of the African Baobab tree, an ancient and mysterious tree that grows in the African savannah. With the gargantuan Baobab tree at the center of the story (it can measure up to 40 feet wide and 60 feet high), we also follow the life cycle of the area's birds, elephants, and plants, which are all supported in some way by the Baobab tree.

Young children especially love the large illustrations of exotic creatures that inhabit the savannah, including the boomslang snake, praying mantis, flap-eared chameleon, and yellow-billed hornbill. This book's tone expresses a deep appreciation of the interconnections among the tree and its inhabitants, and a reverence for all forms of life on the savannah.

Teaching Tips

Tree of Life provides a good starting place for a conversation about ecosystems. Research shows that children tend to grasp ecosystem concepts by the sixth grade. Yet my own experience tells me that they can grasp the interconnected nature of life at much younger ages. During a pilot study that I conducted with a group of fifth graders, we talked about ecosystems. One student, "Shawna," said that her family was like an ecosystem: When one person is sick, it affects everyone else; when one was late for school, they all have to deal with it.

During this study, Shawna and her classmate, a young boy we'll call Dwayne, had this lively exchange:

> **Dwayne:** "I saw it in the [Walt Disney] movie ***The Lion King.*** When you said, 'They are all connected,' I said, 'Yeah, like in the circle of life.'"
>
> **Shawna:** "Yeah, I think the circle of life is basically, it's sort of gruesome, but it's like the food chain [laughing]: An animal eats an animal, eats an animal, and so on and so forth."

Here Dwayne sees a distinction between the linear structure of the food chain and the circular feedback processes that we had been talking about during class:

> **Dwayne:** "And the father Mufasa, he said, 'When we die, our bodies become the grass and the antelopes eat the grass.' And then Chuma says, 'Don't we eat the antelope?' Mufasa says, 'Yes, we are all part of the circle of life.'"
>
> **Shawna** (impressed): "Whoah! Dude!"

Dwayne, who was 10, clearly made the link between the ecosystem metaphor of "the circle of life," Shawna's description of a family as a system, and the abstract conversation we were having about circular feedback loops. Clearly, Walt Disney is masterful at "edu-tainment": creating entertaining stories that often have educational value as well. This encounter made me wonder: Is it possible to harness Hollywood-style entertainment to convey systems concepts to students? I hope so. So stay tuned.

You can also talk about the subtle nature of delays and time horizons while reading this story. As we saw earlier, a time horizon is the interval of time over which a system demonstrates the full pattern of behavior that concerns us. A good practice when you're studying a system's behavior is to ask: "What time frame am I most interested in? One week? One year? Ten years?" For example, if you're thinking about petroleum in a one-year time horizon, you might focus on price and supply. But if you use a 200-year time horizon, you'll likely start thinking about pollution and possible alternatives to petroleum.

Questions to Consider

For Younger Readers

- What is this story about? What happens?
- Is there a beginning and an ending?
- What happens to the African Baobab?

For Older Readers

- Why do you think they call it "the Tree of Life"?
- How do the African people treat the Baobab tree? (You may want to ask kids to compare this story with how the boy treats the apple tree in Shel Silverstein's classic ***The Giving Tree.)***
- Over what time horizon does the tree's life cycle unfold?
- What factors encourage or discourage us to be attuned to natural cycles?

Partner Stories

For other "circle of life" stories, see ***A River Ran Wild*** by Lynne Cherry (see p. 77), ***The Story of the Hawaiian Sea Turtle*** by Marion Coste (University of Hawaii Press, 1993), and ***Brother Eagle, Sister Sky*** by Chief Seattle (illustrated by Susan Jeffers, Dial Books, 1991).

The Lorax

Author: Dr. Seuss
Publisher: Random House, New York, 1971
Format: Picture book, fiction
Age Range: 4–8, but compelling for children of all ages

Systems Thinking Concepts

Simple interconnectedness, balancing the short term with the long term, the impact of time delays, unintended consequences related to the notion that "bigger is always better," the pitfalls of overusing natural or common resources

A Quick Look at the Story

In this tale, the Once-ler, a mysterious character whom we never see (except for his long, green arms!), tells of how he happened upon a lovely, pristine area (which is hauntingly reminiscent of the South American rain forests). The area had an abundance of lush natural resources, including Truffula Trees, Brown Bar-ba-Loots, Swomee-Swans, and Humming-Fishes. Taken by the brightly colored Truffula Tree tufts ("which smelled like fresh butterfly milk"), Once-ler and his ever-growing family chop them down, build a factory, and start to mass produce Thneeds—curious-looking sweater-like objects that no one really needs. As the precious Truffula trees begin to disappear, so do the forests' inhabitants. The lovable old Lorax, a bushy, orange fellow with an Albert Einstein mustache, speaks for the trees, "for the trees have no tongues." He repeatedly cautions the Once-ler not to overdo it, but his warnings go unheeded.

Finally, the last Truffula Tree is cut down. ("No more trees. No

more Thneeds. No more work to be done.") The Lorax flies from the earth, leaving behind a small pile of rocks engraved with one word: "UNLESS." All seems bleak until the Once-ler reveals that he has saved one Truffula Tree seed. He tosses the seed to a curious child, with the hope that this next generation will protect the environment better than he did.

Teaching Tips

"Watch out!" I remember hearing one of my parents say as I was about to order the largest banana split on the menu. "Your eyes may be bigger than your stomach!" The lesson they were trying to help me to learn was that bigger is not always better and, sometimes, such voraciousness has its consequences. (In my case, the lesson involved a half-eaten sundae and, later, a stomachache.)

Most young children will immediately get this story's message about greed. It may take a few readings to explore this book's other equally powerful themes: the cost of not seeing simple interconnections (in this case, between humans and our environment); the pitfalls of overuse of the earth's natural resources; the dangers of pollution; and the importance of considering the long-term consequences of meeting short-term needs.

A good starting point in sharing the lessons of this story is the Once-ler's all-consuming greed and his complete ignorance of the connections between his increased production of Thneeds and the decreased health of the environment around him.

In the drawing below, the interlocking causal loops give us a visual way to explain what may be happening in the story:

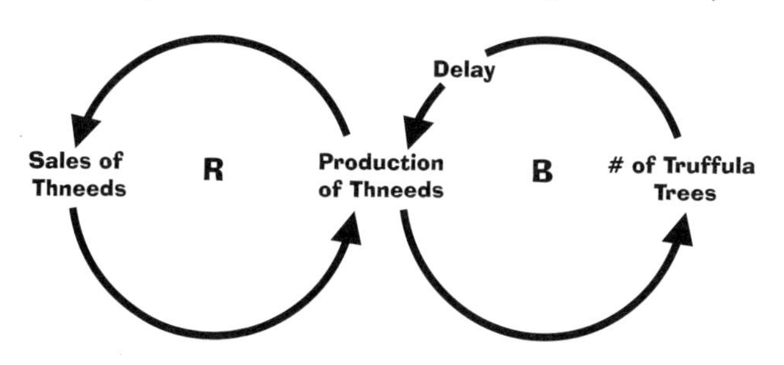

As the sales of Thneeds begins to take off, the Once-ler gets his family to produce more and more Thneeds. As sales and production continue to rise, the number of Truffula Trees begins to decline. But it takes a while (those sneaky delays!) for their decline to become obvious, so the Once-ler keeps clear-cutting the Truffala Tree forest. Eventually, there is only one Truffula Tree left, and the surrounding environment lies in ruins. There are no resources remaining to meet the demands of Thneed production.

The causal loops on p. 89 might well make you think of the "Limits to Success" systems archetype. In this scenario, there are inherent constraints on the Once-ler's behavior that come back to haunt him. The lesson here for everyone is to stop and look twice at what seems like a good idea. Look into the future, think about possible problems and limits that might come up (try drawing them), and then try to *plan* for those problems or limits.

Questions to Consider

For Younger Readers

- What happened in the story?
- What happened to the Truffula Trees?
- Why do you think the Once-ler and his family didn't realize what was happening to the forest? (You can use this as an opportunity to talk about the concept of "delays." For younger kids, you may need to use a word or phrase other than delay; for example, "that waiting time between doing something and seeing what happens.")

For Older Readers

- What was the Once-ler trying to do? What was the Lorax trying to do? How are their hopes similar or different?
- Is the Once-ler thinking more about the past, present, or future? How about the Lorax?
- If you could visit the Once-ler, how would you help him take

better care of his Thneed business? What would you tell him to do, if he could start all over again?

- At what point in the story does the situation get out of control? The beginning, the midpoint, or toward the end? (One idea here is that small, early mistakes are easier to fix than big, later problems.) What choices did the various characters need to make at each point along the way?
- How can we learn to better see what's happening around us? (Another one of those big questions!) By this, I mean it's not always obvious at first what is really happening. Often, by the time we know there is a problem, it is too late to do anything to stop it.

Voices from the Field

Janice C. Kowalczyk, an assistant director of the leadership program in discrete mathematics at Rutgers University, describes her experience using ***The Lorax*** with a group of teachers in a graduate education program:

"I used ***The Lorax*** in this course to bring systems alive in a way that the teachers could connect to and understand. The story approach helped all my teachers at the same time.

We read the first 13 pages aloud and then looked at the relationships involved. For example, we looked at the behavior over time of the trees, the fish, etc., and tried to think about what kind of connections were involved between all the players. We then looked at what factors caused things to change and how and when these come into play.

It gave the teachers the opportunity to build a visual picture or stock and flow model that represents their interpretation of ***The Lorax*** story (see p. 92). Rather than think there is one 'right' model (which is the tendency), asking them to create their own models helped to get across the point that every (systems thinking) model has an underlying story that is a group's or a person's perception of the system they are trying to model....

Using the story helped the teachers to make a connection between what they teach and how a systems perspective can help. My hope is that this will transfer to the way they talk and look at literature as well as other subjects areas with their young students in the future."

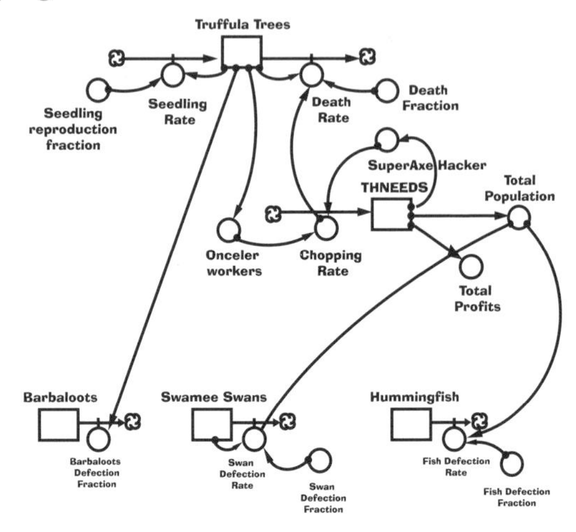

In addition to Janice's innovative use of ***The Lorax,*** several school systems are now using ***The Lorax*** as a way to explore environmental issues and as a basis for understanding stocks and flows.

Partner Stories

Wump World by Bill Peet (Houghton Mifflin Company, 1970), is a fiction storybook for ages 6–10 about environmental abuse and rejuvenation; ***Yertle the Turtle and Other Stories*** by Dr. Seuss (Random House, New York, 1988, pp. 5–10) offers a simple tale about limits to growth; see also ***A River Ran Wild*** by Lynne Cherry (see p. 77), and ***The Wartville Wizard*** by Don Madden (MacMillan Publishing Company,1986).

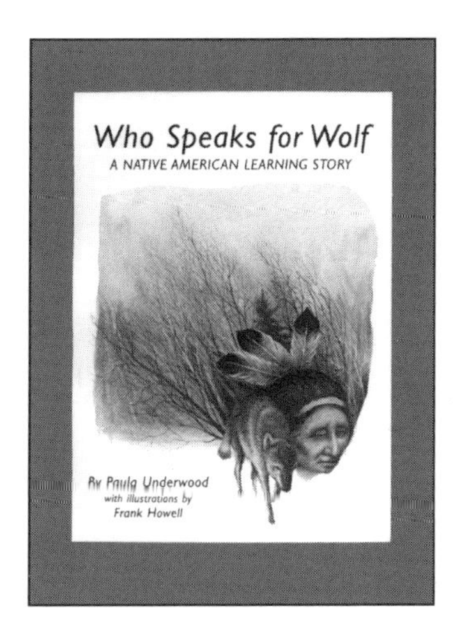

Who Speaks for Wolf? A Native American Learning Story

Author: Paula Underwood, illustrations by Frank Howell
Publisher: A Tribe of Two Press, San Anselmo, California, 1991
Format: Storybook, fiction
Age Range: 3–103!

Systems Thinking Concepts

Simple interconnectedness; unintended long-term consequences; effect of individual seemingly rational behaviors on the "big picture"; multiple causes and effects; the importance of thinking about the whole as well as the parts

A Quick Look at the Story

Paula Underwood, a contemporary Iroquois oral historian, tells this story as it was told to her by her father, Sharp-Eyed Hawk. The story begins with the image and metaphor of a circle: "Almost at the edge of the circle of light cast by central fire—Wolf was standing." (The circle remains a powerful metaphor throughout the story.) An eight-year-old boy asks his grandfather to explain how their family came to live with the wolves. We hear from the grandfather of the dilemmas and unintended consequences that occur when the tribe moves into the wolf community's "Center Place." The story describes the tribe's struggle to live in this new setting as they learn valuable lessons from the wolf community, from nature, and from each other.

Teaching Tips

Who Speaks for Wolf is a poignant tale that encourages readers to start by looking at the big picture and then to see how the various groups are a part of that whole. Through this emphasis on understanding the nature of the whole system, Underwood provides a wonderful balance to the age-old tendency to first think analytically; that is, to treat the whole as something to be taken apart. Children (and adult readers as well) can be encouraged to think through the interconnections that make up the "whole": those connections among the wolf community, the Indian tribe, the arrival of the white settlers, nature, and so on.

Through the wise grandfather, we are also given an elegant model for looking at a problem from many different perspectives: "And so it was / that the People devised among themselves / a way of asking each other questions / whenever a decision was to be made." For children and for grown-ups, the story of ***Wolf*** can help us first to see how we ask questions in troubling situations, and second, to learn how to ask better, more thoughtful questions.

For example, in another related book ***(Three Strands in the Braid: A Guide for Enablers of Learning)***, Underwood describes her father's "Rule of Six": For every perceivable phenomenon, devise at least six plausible explanations. "There will probably be sixty," her father once told her, "but if you devise six, this will sensitize you to the complexity of the Universe, the variability of perception. It will prevent you from fixing on the first plausible explanation as 'the Truth.'" The Rule of Six is a wonderful tool for furthering systems thinking, as it encourages us to consider the perspectives of all. It can also be a fun exercise for children. For example, come up with six reasons why "the food in the school cafeteria is not very good." Or, "Why is a high-school diploma a good thing to have?" Or, "Why do we need to consider everyone's needs?"

Questions to Consider

For Younger Readers

- What did you hear happening in this story?
- What is connected to what or whom? (Encourage your readers to create a map of the whole "system" by naming the interconnections.)
- What might we learn from grandfather's story about Wolf-Looks-at-Fire?

For Older Readers

- Why do you think the community did not listen to Wolf's brother?
- How might the boy and his community have taken the wolf community into account more in their decision?
- What do you think the grandfather is trying to teach the boy by telling this story?
- What do you think was meant by the elder's statement, "Let us now learn to consider wolf?" Can you think of a time in your own experience when someone or some group needed to be represented in a decision-making process, but was not?
- For those teens or adults who are part of an organization, this question may also be useful: Can you think of questions to ask that would better take into consideration the whole system, whether it be your family, your class, or department at work?

Voices from the Field

Mary Ellen Gonzales, story teller and director of the New Mexico LearningWay® Center, tells of her experience using ***Who Speaks for Wolf*** with kindergarten through eighth-grade students:

"I have read and told ***Who Speaks for Wolf*** to students in kindergarten through eighth grade for about eight years now. The story is so all encompassing that I have found it very helpful to decide in advance the direction I want the discussion—which follows the story—to go.

I actually tell ***Wolf*** in two ways—first, I tell it as it is written. I explain that it is a 'Babushka' story. Like the Russian dolls, it is stories within stories. This focuses students' attention more on the literary, technical aspects of the story. That might be appropriate for the class I am telling it to. And, it might not be. The other way I tell ***Wolf*** begins with 'Long ago, long ago … ' Told this way, the story is not as rich, but it is easier to understand; there are not nearly as many elements for students to consider. I recommend starting with this method.

When I tell ***Wolf,*** if I end with, 'And what may we learn from this?' and sit quietly, I nearly always get responses that go something like this: 'It is important to take everyone's perspective into account.' Again, this is so big that most students need a bit of guidance in exploring this concept. I come with a prepared issue to use as an example. The issue can be water restrictions (we have had two very dry years in New Mexico and we do have community-wide water restrictions that affect all of us), wolf reintroduction, a scientific issue, mediation techniques, or whatever I think might be important to the students I am addressing. *It is very important that I have done my homework on the issue I choose to use as an example!* Sometimes, of course, a student will have an issue that he or she wants to consider this way. Then, I ask questions to further my own understanding, not to truly guide the thinking of the students. Both ways are valuable.

Because we as a society do not normally look at any issue systemically, we all (teacher, students, story teller) need to remind each other and ourselves of people, groups, issues, and/or consequences that 'we forgot to consider.' For all of us, it is a learning process; not one of us is an 'expert.' I hope that those moments are moments in which students, and the teacher, and I, can truly learn."

Partner Stories

Similar to ***The Lorax*** by Dr. Seuss (see p. 88), ***Who Speaks for Wolf*** also asks the question "How much was enough and how much was too much?" To learn more about asking generative questions, see ***Three Strands in the Braid: A Guide for Enablers of Learning,*** also by Paula Underwood (A Tribe of Two Press, San Anselmo, CA, 1994). It is an excellent partner book to ***Who Speaks for Wolf.*** You may also want to read ***A River Ran Wild*** by Lynn Cherry (see p. 77) to compare how the different groups handled similar situations.

Paula Underwood died unexpectedly during the writing of this book. She generously read excerpts along the way and made sure I knew that she believed in this little book. In what became her life's work, Paula took on the task of "writing down" the oral history of the Iroquoia people, which we can all learn from today in the form of three learning stories: **Who Speaks for Wolf, Winter White, Summer Gold,** *and* **Many Circles, Many Paths.** *(***Who Speaks for Wolf*** received the Thomas Jefferson Cup Award for quality writing for young people.) As an author, speaker, consultant, and founder of The LearningWay® Center (TLC), she was dedicated to education and cross-cultural understanding and her wisdom was beyond measure. She will be sorely missed.*

Part IV

Endnotes

1. Lorenz mentioned the butterfly example in "Predictability: does the flap of a butterfly's wings in Brazil set off a tornado in Texas?", an address to the annual meeting of the American Association for the Advancement of Science in Washington, December 29, 1979. Also, keep in mind that to say that the butterfly directly "causes" a hurricane would be misleading. We would be just as correct if we said the hurricane was caused by wind that wafted through the butterfly's wings or that brought a scent of pollen to its nose, or by great-Aunt Tilly flapping the laundry. So perhaps one lesson here is for us to become more aware of the exquisite "sensitivity" of complex systems—rather than forcing direct cause-and-effect links when the actual interrelationships are obscured by our imperfect perceptions of the world. Of course, another learning we might take away from the "butterfly effect" story is to be much more humble about our relationship to the "truth."

2. The shape can also be radiating, as when many effects flow from a central cause.

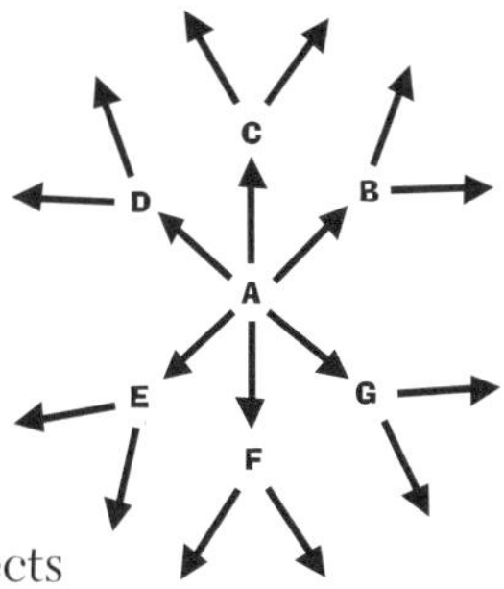

   Or the shape of causality can be branching as well, as in the multiple effects of an oil spill on the environment:

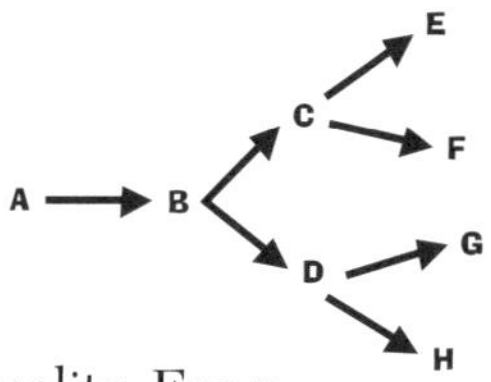

   Educational researcher Tina Grotzer has explored children's understanding of these various models of complex causality. For a fuller discussion of the distinctions made here, see Tina Grotzer's ***Children's Understanding of Complex Causal Relationships in Natural Systems*** (1993, Cambridge:

Harvard Graduate School of Education).

3. For more about "Limits to Success" and the other classic systems archetypes, see the reference booklet ***Systems Archetypes I*** (1992, Waltham: Pegasus Communications) or the workbook ***Systems Archetype Basics*** (1998, Waltham: Pegasus Communications). Both are available through http://www.pegasuscom.com.

4. Donella Meadows, ***The Global Citizen*** (1991, Washington, D.C.: Island Press).

5. During my work with a group of fifth-grade students, I found another interpretation of the performance-practice loop. This one, from Amelia, describes how the element of "fun" influences both performance and the desire to practice:

    Amelia: "See, first you practice a lot and then when your performance gets really good, you don't have to practice a lot but you can't do that if you want to be really, really good and you want to get even better, so you still practice more."

    I: "If you practice more, you play better, is that what you said?"

    Amelia: "Yeah, and then it gets funner, since you are winning . . . then you practice. Then you go back to your practice and then it keeps on going down the circle. So you go, from practice, then to performance, and then how fun it is. And then you keep on going around the circle. And then it gets funner, and you play better, and you practice more, and it keeps on going around the circle!"

6. This book is currently out of print, but may be found at your local library. You can also find copies through an out-of-print search company such as Bookfinders.com.

7. In their book, ***How the Way We Talk Can Change the Way We Work*** (2001, San Francisco: Jossey-Bass), Lisa Lahey and Robert Kegan point to the greater sense of choice, concentration, and power that can come to the human mind through the process of negentropy: "Our bodies are running down, but at the same time, with good luck and supportive efforts, our minds might be 'running up.'"

8. For more on predator/prey relationships and other types of balancing loops, see ***Systems I: An Introduction to Systems Thinking*** (pp. 6–10) by Draper Kauffman, Jr.

9. For a more detailed explanation of the "Shifting the Burden" archetype and additional examples, see ***The Fifth Discipline Fieldbook,*** pp. 138–139.

10. The period during which scientists were monitoring low ozone readings and yet not "seeing" them is described well in Paul Brodeur's article "Annals of Chemistry: In the Face of Doubt," *The New Yorker,* June 9, 1986, p. 71.

11. For a more detailed description of this exercise, see "Frames" in ***The Systems Thinking Playbook, Volume III*** (by Linda Booth Sweeney and Dennis Meadows, Pegasus Communications, 2001).

Guide to Systems Thinking Diagrams

Word-and-arrow diagrams, or what systems thinkers call "causal loop diagrams," are one way of mapping the interrelationships within a system. These diagrams are a visual picture of what we mean when we talk about "structure": the set of interrelationships that drive the patterns of behavior we observe and the events to which we react. Other kinds of diagrams also may show relationships between variables (for example, process flow diagrams or influence diagrams). But what makes causal loop diagrams unique is that they show the movement of circular feedback between variables within a system. That is, they show, for example, that A influences B, which influences C, which comes back and influences A (a cause can become an effect, then become a cause, then become an effect, and so on).

A causal loop diagram consists of variables connected by arrows. The causal links between the arrows are labeled with either a "+" or "–"(or "s" or "o"), to indicate how one variable influences another. Here is a brief overview of the language of links and loops you will encounter as you explore causal loop diagrams further:

+ → S — A causal link between two variables where x adds to y or where x causes a change in y in the same direction.

– → O — A causal link between two variables where x subtracts from y or where x causes a change in y in the opposite direction.

Delay: If there is a significant time delay between two variables, this is represented by drawing two hash marks, or by writing the word "Delay" across the link that connects those two variables.

R A reinforcing feedback loop that amplifies change.

B A balancing feedback loops that seeks equilibrium.

Let's take the example of the flu (previously used by both George Richardson and Daniel Kim), and look at the story described in these loops, beginning on the right side. A few people

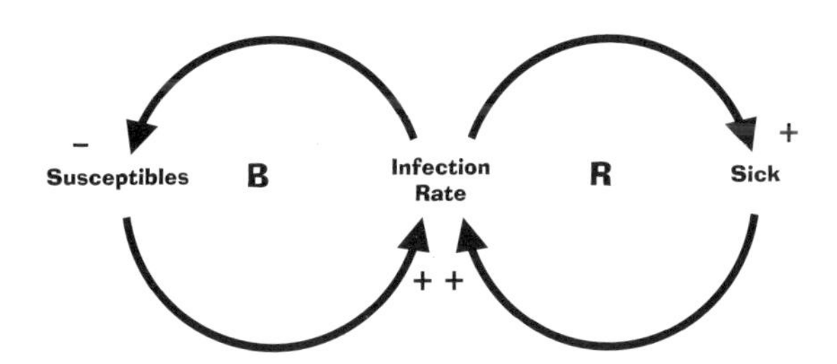

infected with the flu (that is, the number of people who are sick) make contact (therefore increasing the infection rate, and so this is labeled with a + sign) with people who can catch it (those who are susceptible to getting sick). This results in more people becoming sick, so still more "susceptibles" become infected and the actual number of susceptibles declines (thus the – sign). This self-reinforcing process continues until the number of those who are susceptible to getting sick falls low enough to slow down and eventually stop the spread of the flu.

For ideas about how to draw effective causal loop diagrams, see:

"Tips on Drawing Causal Loop Diagrams" (see resource: SE1998-02 Tips for CLDs created by the Catalina Foothills School District. Go to: http://www.clexchange.org).
"Problems in Causal Loop Diagrams Revisited" by George Richardson, The Creative Learning Exchange. ***The Creative Learning Exchange Newsletter:*** Volume 6, Number 3, Fall 1997.

"Guidelines for Drawing Causal Loop Diagrams," a pocket guide of tips published by Pegasus Communications (www.pegasuscom.com).

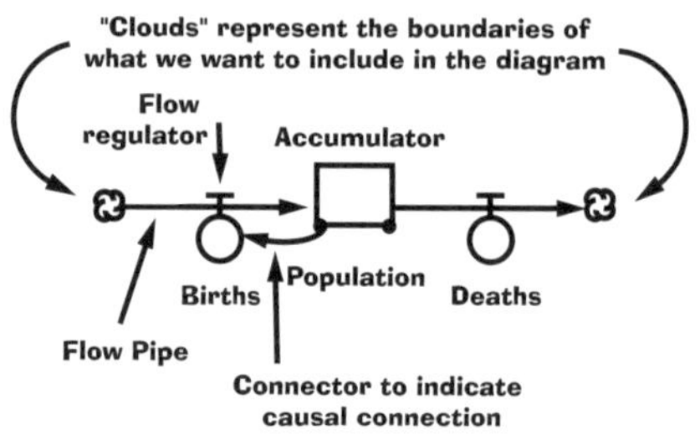

In addition to causal loop diagrams, you'll also discover other kinds of diagrams as you explore the systems thinking world further. These diagrams, consisting of circles, blocks, arrows and cloud-like objects, are called "stock and flow diagrams."

To read one of these diagrams, you first need to know what stocks and flows are. *Stocks* (often called accumulators) are anything that accumulate and that can be measured at one point in time, such as population, the amount of water in a bathtub, and so on. *Flows* represent things that change over time, such as number of births, the inflation rate, etc.

Unlike causal loops, stock and flow diagrams provide information about rates of change. In brief, they show how the various stocks and flows in the system influence one another and how the feedback flows through the system.

These diagrams are also often used to build computer simulation models; the model builder assigns initial values to the stocks (such as "population equals 2,000 at time zero") and rates for the flows (such as "20 births per month"). Also see p. 92 for a sample stock and flow diagram.

Finding Your Own Stories

You, too, can find your own powerful stories for teaching systems thinking ideas and perspectives! To give you some assistance in the process, I suggest you look for the systems concepts described in **Part 1** and for signs of cause-and-effect complexity. You also may want to ask yourself some of the following questions, which I used in the process of developing this book:

- Is there a lot of interconnectedness? By interconnectedness, I mean dynamic situations (those that exhibit change over time) with multiple characters or actions that result in multiple influences or effects. When there are high degrees of interconnectedness, it is often difficult to anticipate the possible consequences of a decision or event. That is what can make the story interesting!
- Does one (or more) character act in an openly aggressive way? Aggression can often indicate an attempt to achieve unlimited growth—which suggests the "Limits to Success" systems archetype mentioned in **Part 1**.
- Does the story feature something or someone acting out of control? If so, this might indicate a reinforcing loop at work.
- Is there a "yo-yo" feeling to the story, where a character is being pulled back and forth in some way? Or is there a situation that seems to get better, then worse, then better again? If so, this might indicate a balancing process.
- What are the elements of surprise or the unintended consequences of the story? Do those consequences feed back to impact the initial causes of the problem or situation? If so, how? Are the characters generally focusing on the short term or the long term? Are some focused on the short term while others emphasize the long term? When characters operate

from different time horizons in this way and pursue separate goals, friction and conflict often arise. The conflict can open a window onto the underlying systems structure driving the characters' behavior in the story.

- Does the overall situation in the story worsen or get better over time? If this happens, the underlying system may be characterized by a dominant balancing loop with delays.

Using This Book with Adults

Although ***When a Butterfly Sneezes*** was designed to be a resource for adults to use with children, many of the stories described in this book can also be used in working with adults as well.

In particular, I've found that some of the stories have universal appeal and are written in a way that is compelling and entertaining to both young readers and grown-ups. These books can be used as fun, non-threatening ways for helping adults to learn about systems thinking. (For more about which books address which systems thinking concepts, see the story overview matrix on pp. 10–11.) So, if you would like to try using stories in working with adults, check out the following titles:

- Any of the Dr. Seuss books. For an example of how a Dr. Seuss book can be used with an adult group, see Don Robadue's "Voices from the Field" on p. 59.

- ***Zoom*** by Istvan Banyai. This book is especially effective for looking at multiple levels of perspective. See Tracy Benson's "Voices from the Field" on p. 75 for an example of how this book was used with a group of educational administrators.

- ***Who Speaks for Wolf? A Native American Learning Story*** by Paula Underwood. I first heard this story during a workshop about organizational learning. Its call for listeners to look at a "problem" from many different perspectives is a helpful reminder for any age group.

Future volumes in this ***Systems Thinking for Kids, Big and***

Small series will focus on books that are especially effective with older readers and adults. Some of the stories featured will include:

- ***Why Mosquitos Buzz in People's Ears: A West African Fairy Tale*** by Verna Aardema
- ***The King's Stilts*** by Dr. Seuss
- ***Billibonk and the Thorn Patch*** by Philip Ramsey
- "Top of the Food Chain" in ***Without a Hero (and Other Stories)*** by T. Coraghessan Boyle
- ***The Wisdom of the Crows and Other Buddhist Tales,*** retold by Sherab Chodzin and Alexander Kohn
- ***Robin Hood*** by Howard Pyle. Try reading this book again with the "Fixes that Fail" archetype in mind!
- ***Hamlet*** and ***Macbeth*** by William Shakespeare can provide wonderful examples of ways to explore the "Limits to Growth" archetype[1].

Tips for Working with Adults

The following tips are designed to help give you guidance in using these or other stories with adults. Before you begin, look in the "Voices from the Field" sections in Part III of this book for more ideas.

Tip 1: Talk About Stories in General

Talk with the group about the structure and the content of the

[1] Thanks to Timothy Joy, an educator at La Salle High School who has generously shared his ideas for looking at several Shakespearian stories from a systems perspective.

Western stories they are familiar with. You may ask, as I have in this book, "Are we, particularly in Western cultures, actually *taught* by the structure of the stories we read to look for linear rather than systemic relationships?"

Tip 2: Read the Story Aloud

Adults, like children, love to hear stories. Some even like to read them aloud. If you have a small group (8–12), try sitting in a circle and passing a copy of a book around, with each person reading one or two pages. If you are working with a large group, you may want to read the book aloud yourself, showing the illustrations to the group on an overhead projector as you go along (make sure you obtain rights from the publisher to duplicate the illustrations for this purpose).

Tip 3: Begin by Asking the Question "What Happened?"

What relationships do your listeners notice in the story, and what happens to those relationships over time? Make sure you leave time and space for people to share their interpretations of the story, or to ask questions. This is particularly important if you have participants from different countries who may need to clarify some of the language or allusions in the story. See the section "Questions to Consider" for older kids with each story description in Part III of this book for some ideas. Many of the questions are appropriate for adults, and you'll probably find that they help to generate lively conversations.

Tip 4: Introduce Selected Systems Thinking Concepts

If your group has some familiarity with systems thinking concepts, they will automatically integrate some of what they already know into the discussion about the story. If not, introduce the appropriate systems thinking concepts as it seems appropriate. If someone begins to describe a causal feedback loop, ask them to draw it on a flipchart or overhead transparency for easy reference. You may also wish to introduce a template for talking about causality; for ideas, see Tip 3: "Introduce the Language of Causality" on p. 41.

These last three tips come from Diane Cory, organizational learning consultant and storyteller:

Tip 5: Respect the Stories and the People Who Are Listening

For effective storytelling with adults, it is important that you personally respect the process of telling, the role you play in the telling, and also the people who are listening. It's also imperative that you understand you are giving voice to the stories; that they are using you to serve a greater wholeness.

Tip 6: Listen to Your Mind and to Your Whole Body for Cues

Listen to how your mind and body can give you the correct timing for storytelling—the right moment to share a story. Allow yourself to *feel* your way into the telling. Get down and crawl on your mental hands and knees if you need to. You'll be less likely to trip yourself up and fall.

Tip 7: Love the Stories

Love the people you are telling them to. Love the moment you are telling them in. Said another way, enjoy telling the stories as much as you expect the group to enjoy hearing it.

Additional Resources

There's a variety of additional resources available for helping you explore further the field of systems thinking as well as how to teach and learn through stories. The following are some books, videos, and URLs that you may find useful.

Learning Through Stories

Call of Stories: Stories and the Moral Imagination by Robert Coles. In this lovely book, psychiatrist and educator Robert Coles looks at the role that stories and storytelling play in lives of children and the potential these stories have to teach moral values (1989, Boston: Houghton Mifflin Company).

"*Strategic Stories: How 3M Is Rewriting Business Planning*" by G. Shaw, R. Brown, et al. (*The Harvard Business Review*, May–June 1998). This short but powerful article describes the use of "strategic storytelling" at 3M (the inventor of Post-It notes). The authors recognize that people tend to default to using bullet points in business communication, which maximizes speed and clarity, but at the expense of other elements such as appreciating causality and effectively communicating dynamic complexity.

Teaching as Story Telling: An Alternative Approach to Teaching & Curriculum in the Elementary School by Kieran Egan. This book offers some great examples of how to use stories in order to teach any subject in a more engaging and meaningful way (1986, Ontario: The Althouse Press).

Helping Children Develop Systems Thinking Skills

Building a Global Civic Culture: Education for an Interdependent World by Elise Boulding. In this inspirational book, Boulding uses many systems thinking concepts to help us think about building a global civic culture in a world in which we "share a common space, common resources, and common opportunities" (1988, Syracuse: Syracuse University Press).

The Creative Learning Exchange (CLE) is the source for those wanting information on introducing systems thinking (particularly system dynamics) in kindergarten through 12th grade education. CLE is a nonprofit foundation that acts as a clearinghouse to provide information such as curriculum units and articles to help teachers and practitioners share their experiences. This web site lists innovative curriculum using stories to bring teaching systems concepts to life in educational settings. 1 Keefe Road, Acton, MA 01720; tel: 508-287-0070; fax: 508-287-0080; web: www.clexchange.org.

Schools That Learn: A Fifth Discipline Fieldbook for Educators, Parents and Everyone Who Cares About Education by Peter Senge, Nelda Cambron-McCabe, Timothy Lucas, Bryan Smith, Janis Dutton, and Art Kleiner. This book provides a terrific summary of systems thinking as it is being used in K–12 education in the U.S. (2000, New York: Currency/Doubleday).

Systems One by Draper Kaufmann, Jr. (1980, Minneapolis: Carlton, Publisher). This is one of the best introductions to systems thinking ever written. It is full of examples and fun illustrations, suitable for sharing with older children. (Available through Pegasus Communications, Inc., 1-800-272-0945 or www.pegasuscom.com.)

The Systems Thinking Playbook, Vols. 1–3, by Linda Booth Sweeney and Dennis Meadows. This book is full of fun, experiential exercises designed to help people learn about systems thinking ideas. It has been used both by adults working with adults and by educators working with children (Revised and expanded edition,

2001, Waltham: Pegasus Communications, Inc.)

Termites, Turtles and Traffic Jams: Explorations in Massively Parallel Microworlds by Mitchel Resnick. If you're interested in the notion of complex systems, both natural and social, then this is a good book to read. Resnick explains his methods for teaching high school students about the world of decentralized systems—systems in which individuals with no centralized authority (birds in a flock, the immune system, ant colonies, cars on a highway, market economies) follow simple rules and cause patterns to emerge via their interactions (1995, Cambridge: The MIT Press).

Toward Systemic Education of Systems Scientists by Russell Ackoff and J. Gharajedaghi. The authors describe what they call "machine age education." Ackoff says "schools are modeled after factories," and describes the shift that needs to occur in order to realize "systems age" education (1985, *Systems Research,* Vol. 2, No. 1, p. 21–27).

Systems Thinking for Grown-Ups

Business Dynamics: Systems Thinking and Modeling for a Complex World by John Sterman. This textbook is for those who are serious about learning how to apply system dynamics concepts and tools in real-world situations. While it is hefty (1008 pages), it is full of wonderful, clear, and often humorous examples that make the concepts come to life (2000, New York: Irwin/McGraw-Hill.)

Feedback Thought in Social Science and Systems Theory by George Richardson. If "systems thinking" sounds like a fad to you, then read this book. Richardson traces the history of key systems concepts (i.e. feedback) from ancient times to the present. Richardson has a wonderful talent for making the somewhat abstract concepts of systems thinking come alive. I recommend reading any articles you can find that are authored by him (see the CLE website for references to his articles and videos). (1991, Waltham: Pegasus Communications, Inc.)

The Fifth Discipline Series—Peter Senge published ***The Fifth Discipline: The Art and Practice of the Learning Organization*** in 1990. This book sparked a revolution in the way we look at organizational learning and at systems—resulting in a series of additional resources. In addition to ***Schools That Learn*** (see above), Senge has convened seasoned practitioners to co-author ***The Fifth Discipline Fieldbook*** and ***The Dance of Change: The Challenge of Sustaining Momentum in Learning Organizations*** (1997/1999, New York: Doubleday).

Global Citizen by Donella Meadows. In this book, Meadows give us page after page of accessible facts about the impact of human behavior on the environment (1991, Washington D.C.: Island Press). Meadows also publishes the ***"Global Citizen,"*** a bi-weekly newspaper column, which can be found through
http://iisd1.iisd.ca/pcdf/meadows/.

Introduction to Systems Thinking by Daniel Kim. A clear explanation of the basic terms, concepts and tools associated with systems thinking (1999, Waltham: Pegasus Communications).

The Logic of Failure by Dietrich Dörner. If you need to be convinced that systems thinking ideas and tools apply in the "real world," take a look at this book. In it, Dörner describes the nature of many historical and current catastrophes in which seemingly sensible steps set the stage for future disasters (1996, Perseus Press).

Only Connect! An Annotated Bibliography Reflecting the Breadth and Diversity of Systems Thinking by David Lane and Michael Jackson. If you are hooked on some of the simple ideas in ***When a Butterfly Sneezes,*** you may want to check out this article to see how systems thinking has been and is being used around the world. (1995, *Systems Research,* (12). 3, 217-228). For a more in-depth look at a variety of systems thinking practices, you may also want to read ***Creative Problem Solving*** by Robert Flood and Michael Jackson (1991, Chichester: John Wiley & Sons).

Powers of 10 video, by Charles and Ray Eames. This short film is thought-provoking for grown-ups and older children alike. It is an excellent way to explore the notion of "levels of perspective" and to see ourselves as living within a series of nested systems (***The Films of Charles and Ray Eames: Volume 1: Powers of 10.*** Pyramid Home Video, Santa Monica, CA. 1-800-421-2304).

Seeing Nature: Deliberate Encounters with the Natural World by Paul Krafel. This lovely book is full of true stories about coming to understand the many interrelationships in the natural and human-made world. (1999, Chelsea Green).

Seeing Systems: Unlocking the Mysteries of Organizational Life by Barry Oshry. The first few pages are not to be missed for an engaging, clear and compelling introduction to systems thinking ideas. This resource is particularly good for those who want to understand the application of systems thinking in organizational contexts (1996, San Francisco: Berrett-Koehler).

Systems Archetypes I is a short booklet that outlines the basic structure and storyline of each of the archetypes (1992, Waltham: Pegasus Communications, Inc.). For practice exercises and more in-depth looks at each archetype and their patterns of behavior over time, you can also check out the workbook ***Systems Archetype Basics*** (1998, Waltham: Pegasus Communications, Inc.)

The System Dynamics Review (Summer 1993). This entire issue was devoted to systems thinking in education. It is available from the System Dynamics Society at http://www.albany.edu/cpr/sds.

The Web of Life: A New Scientific Understanding of Living Systems by Fritjof Capra. If you have been overwhelmed by the "new sciences," you can relax. In this book, Capra gives us an eloquent and respectful synthesis of chaos theory, complexity theory, Gaia theory, and systems theory. (1996, New York: Doubleday).

Related Web Sites

For additional resources related to systems thinking and organizational learning, see http://www.pegasuscom.com.

For a self-study guide to system dynamics called ***"Road Maps"*** (available for downloading), see http://sysdyn.mit.edu

For an example of systems thinking and system dynamics integrated into K–12 curricula, see the Waters Grant project site at http://www.watersfoundation.org

For a short, non-technical introduction to systems thinking, check out http://www.thinking.net/Systems_Thinking/systems_thinking.html

When a Butterfly Sneezes . . . Staying Connected and Sharing Ideas

We have set up an online discussion forum for those who are interested in connecting with others who have read this book. Please visit the discussion group to share your own experiences using the stories in this volume, or to learn more about how to use stories to explore systems thinking ideas. You can also exchange ideas about other books for younger readers, books for older readers (i.e. chapter books and novels, which is the focus of the next volume) or about your work with systems thinking! Also, visit this site to sign our ***When a Butterfly Sneezes*** reading circle guestbook, so we can notify you when future volumes in the series are published.

http://www.pegasuscom.com/butterfly.html.

If you have comments, story suggestions, or questions you would like to share with the author, please do contact Linda by email at **Linda_Booth_Sweeney@harvard.edu.**

Credits

Book cover from ***If You Give a Mouse a Cookie,*** Copyright © 1985 by Laura Joffe Numeroff, Felicia Bond (illustrator). Reprinted by permission of HarperCollins Juvenile Books.

Book cover from ***The Old Ladies Who Liked Cats,*** Copyright © 1991 by Carol Greene, Loretta Krupinski (illustraator). Reprinted by permission of HarperCollins Juvenile Books.

From ***The Cat in the Hat Comes Back*** by Dr. Seuss, TM and copyright © 1958, renewed 1986 by Dr. Seuss Enterprises, L.P. Used by permission of Random House Children's Books, a division of Random House, Inc.

Cover reprinted with the permission of Atheneum Books for Young Readers, an imprint of Simon & Schuster Children's Division from ***Once a Mouse: A Fable Cut in Wood*** by Marcia Brown. Copyright © 1961 Marcia Brown.

From ***The Sneetches and Other Stories*** by Dr. Seuss, TM and copyright © by Dr. Seuss Enterprises, L.P. 1953, 1954, 1961, renewed 1989. Used by permission of Random House Children's Books, a division of Random House, Inc.

"Cover," from ***Anno's Magic Seeds*** by Mitsumasa Anno, copyright © 1992 by Kuso-Kobo. Translation copyright © 1995 by Philomel Books. Used by permission of Philomel Books, an imprint of Penguin Putnam Books for Young Readers, a division of Penguin Putnam, Inc.

"Cover," from ***Zoom*** by Istvan Banyai, copyright © 1995 by Istvan Banyai. Used by permission of Viking Penguin, an imprint of Penguin Putnam Books for Young Readers, a division of Penguin Putnam, Inc.

Book cover from ***A River Ran Wild,*** Copyright © 1992 by Lynne Cherry, reprinted by permission of Harcourt, Inc.

From ***The Butter Battle Book*** by Dr. Seuss, TM and copyright © by Dr. Seuss Enterprises, L.P. 1984. Used by permission of Random House Children's Books, a division of Random House, Inc.

From ***Tree of Life*** by Barbara Bash. Copyright © 1989 by Barbara Bash. By permission of Little, Brown and Company (Inc.).

From ***The Lorax*** by Dr. Seuss, TM and copyright © by Dr. Seuss Enterprises, L.P. 1971, renewed 1999. Used by permission of Random House Children's Books, a division of Random House, Inc.

Cover from ***Who Speaks for Wolf*** by Paula Underwood. Copyright © 1991 by Paula Underwood. Reprinted with permission from A Tribe of Two Press.

About the Author

Linda Booth Sweeney is an educator, researcher, and speaker dedicated to helping children and adults understand how the mysterious natural and social worlds function through the field of systems thinking. She is co-author of ***The Systems Thinking Playbook, Volumes 1-3*** (with Dennis Meadows, Turning Point, 1995/1996/2000) and author of numerous articles on systems thinking.

Linda received her master's degree in human development and psychology from the Harvard Graduate School of Education and is currently completing her doctoral studies there. As a researcher and consultant, she has worked with the Organizational Learning Center at M.I.T. (now SoL, the Society for Organizational Learning) over the last eight years. In that role, she has helped member companies to develop capacity in systems thinking and its related disciplines. She has worked with local as well as international clients, including those in New Zealand, Europe, and South Africa.

She is currently a research member of Harvard's Change Leadership Group, a Gates Foundation–funded research organization focused on developing practical knowledge and knowledgeable practitioners to assist school leaders in their efforts to improve all students' achievement.

Linda lives in Cambridge, Massachusetts with her husband John and two-year old son. She is currently at work on several "systems-oriented" stories for children, while eagerly awaiting the arrival of their second child in January 2001.